Drought, Risk Management, and Policy

Decision Making under Uncertainty

Drought, Risk Management, and Policy
Decision Making under Uncertainty

Edited by Linda Courtenay Botterill and Geoff Cockfield

CRC Press
Taylor & Francis Group
Boca Raton London New York

CRC Press is an imprint of the
Taylor & Francis Group, an **informa** business

CRC Press
Taylor & Francis Group
6000 Broken Sound Parkway NW, Suite 300
Boca Raton, FL 33487-2742

First issued in paperback 2017

© 2013 by Taylor & Francis Group, LLC
CRC Press is an imprint of Taylor & Francis Group, an Informa business

No claim to original U.S. Government works

ISBN-13: 978-1-4398-7629-9 (hbk)
ISBN-13: 978-1-138-07389-0 (pbk)

Library of Congress Cataloging-in-Publication Data

Drought, risk management, and policy : decision making under uncertainty / editors, Linda Courtenay Botterill and Geoff Cockfield.
    p. cm. -- (Drought and water crises in the 21st century ; 2)
    Includes bibliographical references and index.
    ISBN 978-1-4398-7629-9 (alk. paper)
    1. Drought relief--United States. 2. Drought relief--Australia. 3. Environmental policy--United States. 4. Environmental policy--Australia. 5. Climatic changes--United States. 6. Climatic changes--Australia. I. Botterill, Linda Courtenay. II. Cockfield, Geoff.

HV626.U6D764 2013
363.34'9295610973--dc23
                          2012031228

Visit the Taylor & Francis Web site at
http://www.taylorandfrancis.com

and the CRC Press Web site at
http://www.crcpress.com

Contents

Section III   From Policy to Action

Series Editor's Preface

In 2005, I edited a book for CRC Press, *Drought and Water Crises: Science, Technology, and Management Issues.* This new book series by the same name is intended to expand on the theme of the original book by providing new information and innovative approaches to drought monitoring and early warning systems, mitigation, planning, and policy and the linkages between these challenges and important natural resources and environmental issues such as climate change, including increased climate variability, water scarcity, food security, desertification, transboundary water-related conflicts, and water management, to name just a few. There is an increasing demand for more information from scientists, natural resource managers, and policy makers on issues related to these challenges that are at the intersection of drought and water management issues as pressure on the world's finite water resources intensifies.

Drought is a normal part of the climate for virtually all climatic regimes. It is a complex, slow-onset phenomenon that affects more people than any other natural hazard and results in serious economic, social, and environmental impacts. Drought affects both developing and developed countries, but in substantially different ways. This fact has been exemplified in recent years by the severe droughts in the Greater Horn of Africa, China, India, Australia, and the United States. Society's ability to manage droughts more effectively in the future is contingent upon a paradigm shift—moving from a crisis management to a risk-based management approach directed at increasing the coping capacity or resilience of nations to deal effectively with extended periods of water shortage.

At this writing, only one nation, Australia, has a national drought policy in place that is aimed at drought risk reduction. However, there are significant international initiatives underway with the goal of motivating all drought-prone nations to develop national drought policies. Leadership for these efforts is being provided by several key agencies of the United Nations, including the World Meteorological Organization, U.N. Convention to Combat Desertification, and the Food and Agriculture Organization. This effort, coupled with the actions of many other governmental and non-governmental organizations, will, hopefully, stimulate many more nations to move toward the development of national drought policies. *Drought, Risk Management, and Policy: Decision Making under Uncertainty* contributes in a substantial way to this discussion and the ensuing debate on the merits of national drought policy over the traditional crisis management approach.

Linda Botterill and Geoff Cockfield are eminently qualified as editors of this volume through their extensive experience with the policy process in Australia and elsewhere. I have collaborated with Linda on national drought policy research issues for the past ten years and Geoff has a broad experience in agricultural policy. Together they have assembled a series of important contributions to the discussion of drought policy, comparing the experiences of Australia with the process that has unfolded in the United States over the past decade.

This is the second book to be published in this book series and it will certainly heighten awareness of the process of drought policy development as part of broader discussions on preparing for an uncertain future climate.

Donald A. Wilhite
University of Nebraska-Lincoln, U.S.A.

Preface

The relationship between science—and indeed all forms of expertise—and policy development has been the subject of a wide range of research from the science and technology studies tradition, which is strongly sociological in orientation, to the knowledge utilisation literature associated with policy sciences/public policy. Science/policy relationships can be problematic, particularly in democracies where policy making is somewhat of a balancing act between competing societal values. In open and competitive policy communities, much to the frustration of some scientists, expert advice is only one voice around the policy table and often is not decisive in the decision-making process. Yet aspirations to bring order and evidence to policy processes persist, so as Harold Lasswell (1951, 3) asked over 60 years ago,

> What are the most promising methods of gathering facts and interpreting their significance for policy? How can facts and interpretations be made effective in the decision-making process itself?

Policy makers clearly want to base their decisions on the best available evidence and researchers in turn seek to influence policy. However, the interface between science and policy is not an easy one as the imperatives of research-based science and those driving government decision makers are often quite different and there is a level of ignorance on the part of the inhabitants of the 'two worlds' of the needs and constraints of each other.

The editors of this collection have been involved in making, studying, and critiquing Australia's National Drought Policy since its inception in 1992 and our contributors have also been engaged in a variety of capacities with drought sciences and policy in Australia and the United States. This book is one of the outputs of a project supported under the Australian Research Council's Discovery Projects funding scheme (project number DP1096759). As part of this project, the chief investigator, Linda Botterill, received an International Collaboration Award, which funded her visit to the United States in 2010 to undertake research on *the role of science in the policy process: responding to drought in Australia and the United States*. During that period she was hosted by the National Drought Mitigation Center at the University of Nebraska-Lincoln and she would like to acknowledge their generosity during her time in Lincoln.

This collection builds on research undertaken in the United States and, on return to Australia, including interviews with Australian drought scientists and policy makers and the survey of farmer attitudes to drought science and policy reported in Chapter 12. A workshop was held in Canberra in May 2011 to discuss the major themes of the book and included a discussion with officials from Australian government agencies to get their perspective on the issues being addressed by our authors. Pulling together a collection such as this, which straddles not only international borders but also disciplinary boundaries, is a challenge. We would like to thank our publishers for their patience with the production of this book and also that of our contributors, who exhibited considerable forbearance with our requests for refocusing of their chapters.

Linda Botterill and Geoff Cockfield, editors

Reference

Lasswell, H. D. 1951. The policy orientation. In *The policy sciences,* eds. D. Lerner and H. D. Lasswell, 3–15. Stanford, CA: Stanford University Press.

Editors

Dr Linda Botterill is Professor in Australian Public Policy in the Faculty of Business, Government & Law at the University of Canberra. Her research interest is the policy development process, particularly the role of values in the policy process with a focus on rural policy. She is widely published in public policy and political science journals and is the author of *Wheat Marketing in Transition: The Transformation of the Australian Wheat Board* (Springer 2012) and the coeditor of *The National Party: Prospects for the Great Survivors* (Allen & Unwin 2009, with Geoff Cockfield), *From Disaster Response to Risk Management: Australia's National Drought Policy* (Springer 2006, with Donald A. Wilhite), and *Beyond Drought: People, Policy and Perspectives* (CSIRO Publishing 2003, with Melanie Fisher).

Geoff Cockfield is Associate Professor in Politics and Economics and a research associate in the Australian Centre for Sustainable Catchments at the University of Southern Queensland. Geoff worked in agricultural industries and rural journalism before starting an academic career. He has undertaken research on sustainable grazing practices, the socioeconomics of forest vulnerability under climate change, and the economic impact of changes in water availability in the Murray–Darling Basin. With Linda Botterill, he has edited a book on the National Party of Australia and produced a number of articles on agricultural policy in Australia.

Contributors

Lisa S. Darby is a research meteorologist in the National Integrated Drought Information System (NIDIS) Program Office and is based at the National Oceanic and Atmospheric Administration's Earth System Research Laboratory, 325 Broadway, R/PSD, Boulder, Colorado, 80305; e-mail: lisa. darby@noaa.gov. She is currently coleading (along with C. McNutt) the NIDIS Apalachicola–Chattahoochee–Flint River Basin Pilot.

Professor Stephen Dovers is Director of the Fenner School of Environment and Society, Australian National University, and an ANU Public Policy Fellow. His research and teaching interests are in environmental policy and institutions, climate adaptation, and disaster management. Among his more than two hundred research publications are the books *Institutional Change for Sustainable Development* (Edward Elgar 2004, with R. Connor), *Environment and Sustainability Policy* (Federation Press 2005). and *The Handbook of Disaster and Emergency Policies and Institutions* (Earthscan 2007, with J. Handmer).

Tonya R. Haigh is a rural sociologist with the National Drought Mitigation Center (NDMC), located at the University of Nebraska-Lincoln. She has worked with Great Plains ranchers and range advisors to develop Managing Drought Risk on the Ranch, an online drought planning and management resource located at www.drought.unl.edu/ranchplan. She has also coordinated sustainable agriculture education and advocacy efforts throughout the northern Great Plains. Her research interests include climate-related adaptive capacity for agriculture and rural communities and the role of climate information technology and dissemination in increasing adaptive capacity.

Dr Michael J. Hayes is currently the Director for the National Drought Mitigation Center (NDMC) and Professor in the School of Natural Resources at the University of Nebraska-Lincoln. He became the NDMC's director in August 2007 and has worked at the NDMC since 1995. The NDMC now has 16 faculty and staff working on local, tribal, state, national, and international drought-, climate-, and water-related issues. Dr Hayes's main interests deal with drought monitoring, planning, and mitigation strategies. Dr Hayes received a bachelors degree in meteorology from the University of Wisconsin-Madison, and his masters and doctoral degrees in atmospheric sciences from the University of Missouri-Columbia.

Dr Peter Hayman is the principal scientist in climate applications at the South Australian Research and Development Institute (SARDI) based at the Waite Institute, a position he has held since May 2004. Prior to moving to

Adelaide he was coordinator of climate applications for NSW Agriculture. He is an agricultural scientist with an interest in applying climate information to dryland and irrigated farming systems with a recent focus on impacts and adaptation to climate change in the irrigated wine grape industry and low rainfall grains industry. In 2010 Peter Hayman was appointed part of the leadership group of the Primary Industries Adaptation Research Network hosted by Melbourne University and funded by the National Climate Change Adaptation Research Facility.

The Honourable **John Kerin,** AM, is an economist and former politician who is chair of the Crawford Fund, which promotes and funds international research and development activities involving Australian researchers. He was elected to the Australian Parliament in 1972 and was Minister for Primary Industries (1983–1991), Transport (1991), and Trade and Overseas Development (1992–1993) and was Treasurer (1991). Before his election to the federal Parliament he worked at the Australian Bureau of Agricultural and Resource Economics. Since leaving politics he has served on a number of agricultural and natural resources research organisation boards, including the CSIRO.

Dr Cody L. Knutson is a Research Associate Professor with the National Drought Mitigation Center (NDMC) at the University of Nebraska-Lincoln. At the NDMC, he is the leader of the Planning and Social Science Program Area. Since 1997, his primary role has been helping individuals, communities, tribes, states, and national governments prepare for and respond to drought. Dr Knutson's specialty is investigating how regions and activities are vulnerable to drought and collaborating with stakeholders to develop strategies to reduce drought's negative effects. Dr Knutson has served on subcommittees of the US Western Governor's Association's Western Drought Coordination Council, US National Drought Policy Commission, and the United Nations International Strategy for Disaster Reduction's Drought Advisory Group.

Chad A. McNutt is the National Integrated Drought Information System (NIDIS) deputy program manager and is based at the National Oceanic and Atmospheric Administration's Earth System Research Laboratory, 325 Broadway, R/PSD, Boulder, Colorado, 80305; e-mail: chad.mcnutt@noaa.gov. He is currently coleading (along with L. Darby) the NIDIS Apalachicola–Chattahoochee–Flint River Basin NIDIS Pilot.

Roger S. Pulwarty is the Director of the National Integrated Drought Information System (NIDIS) and is the chief of the Climate and Societal Interactions Program at the National Oceanic and Atmospheric Administration (NOAA). He is based at NOAA's Earth System Research Laboratory, 325 Broadway, R/PSD, Boulder, Colorado, 80305, USA; e-mail: roger.pulwarty@noaa.gov. His research interests are in climate variability and change in the western United States, Latin America, and the Caribbean;

assessing social and environmental vulnerability; and designing climate services to meet information needs in water resources, ecosystems, disasters, and agricultural management.

Lauren Rickards is a research fellow at the University of Melbourne, where she works with a range of groups, including the Primary Industries Adaptation Research Network and the Melbourne Sustainable Society Institute. A Rhodes Scholar, Lauren has an interdisciplinary background in cultural geography and ecology, which she has used to interrogate numerous strategic and conceptual issues in agriculture, with a focus on climate variability and change. As both a private sector consultant and academic, Lauren has conducted in-depth longitudinal ethnographic research into farming communities' perceptions and experiences of drought, as well as social research into researchers' perceptions and experiences of doing interdisciplinary and transdisciplinary research. In 2009, Lauren won the excellence in extension (young professional) award from the Asia-Pacific Extension Network in recognition of her intellectual contribution to the field. She is currently deputy chair of the Australian government's Terrestrial Ecosystem Research Network.

Professor Daniela Stehlik is Professor of Sociology and previously foundation Director of The Northern Institute at Charles Darwin University, Northern Territory. She is an adjunct professor at James Cook University (Cairns) and the Australian National University (Canberra). Professor Stehlik is currently Chair, Rural Industries Research and Development Corporation and was a member of the Minister for Sustainable Population's Advisory Panel on Demographic Change and Liveability. In 2008 Professor Stehlik was a member of the Expert Social Panel involved in the federal government's review of drought policy.

Professor Roger Stone is Director of the Australian Centre for Sustainable Catchments and professor in climatology in water resources at the University of Southern Queensland. He also holds the open program chair within the United Nations-WMO Commission for Agricultural Meteorology to provide research leadership globally in the field of climate variability, climate change, and natural disasters in agriculture. Professor Stone is also an expert team leader within the UN Commission for Climatology, Geneva. Professor Stone has continued to publish widely in the international scientific literature and has led climate applications modelling research and development programs, especially those that integrate statistical/mathematical climate models, agricultural models, and decision systems that have direct relevance to industry.

James P. Verdin is the Deputy Director of the National Integrated Drought Information System (NIDIS) and is also the US Geological Survey Famine Early Warning Systems Network program manager. He is based at the

National Oceanic and Atmospheric Administration's Earth System Research Laboratory, 325 Broadway, R/PSD, Boulder, CO, 80305; e-mail: verdin@usgs.gov. He is also leading the NIDIS Upper Colorado River Basin Pilot.

Dr Gene Whitney served as the head of the Energy and Minerals Section of the Congressional Research Service at the Library of Congress advising members of the US Senate and House of Representatives and their staffs on a range of energy issues. Formerly, Dr Whitney was a research scientist and manager at the US Geological Survey and served for 7 years in the White House Office of Science and Technology Policy as Assistant Director for Environment. His areas of responsibility within the Executive Office of the President included science and technology policy of natural hazards and disasters, energy, climate change, water, environment, earth sciences, and earth observations. Dr Whitney holds a BS, MS, and PhD in geology.

Glossary

ABARE: Australian Bureau of Agricultural and Resource Economics

ABARES: Australian Bureau of Agricultural and Resource Economics and Sciences

ACANZ: Agriculture Council of Australia and New Zealand

ACF: Apalachicola–Chattahoochee–Flint

ACF-S: ACF stakeholders

ANU: Australian National University

APSC: Australian Public Service Commission

APSRU: Agricultural Production Systems Research Unit

ARMCANZ: Agriculture and Resource Management Council of Australia and New Zealand

BAE: Bureau of Agricultural Economics (Australia)

BOM: Bureau of Meteorology (Australia)

BRS: Bureau of Rural Sciences (Australia)

BSE: Bovine spongiform encephalopathy

CCC: Colorado Climate Center

CRB: Colorado River Basin

CRCs: Cooperative Research Centres (Australia)

CRTF: Homeland Security Advisory Council's Community Resilience Task Force (US)

CSIRO: Commonwealth Scientific and Industrial Research Organisation (Australia)

DAFF: Department of Agriculture, Fisheries and Forestry (Australia)

DHS: Department of Homeland Security (US)

DIR: US Drought Impact Reporter, www.droughtreporter.unl.edu

EC: Exceptional circumstances

ENSO: El Niño Southern Oscillation

EU: European Union

FEMA: Federal Emergency Management Agency (US)

FEWS: Famine early warning system

FEWS NET: Famine early warning systems network

GAO: US Government Accountability Office, http://www.gao.gov/

HILDA: Household, Income and Labour Dynamics in Australia survey

HSPD-8: Homeland Security Presidential Directive 8 (US)

MDB: Murray–Darling Basin (Australia)

NAMS: National Agricultural Management System (Australia)

NCDC: National Climatic Data Center (US)

NDMC: National Drought Mitigation Center, University of Nebraska, Lincoln, www.drought.unl.edu

NDP: National Drought Policy (Australia)

NDPC: National Drought Policy Commission (US)
NDRA: Natural disaster relief arrangements (Australia)
NIDIS: National Integrated Drought Information System (US)
NOAA: National Oceanic and Atmospheric Administration (US)
NRAC: National Rural Advisory Council (Australia)
NRCS: USDA Natural Resources Conservation Service
NRM: Natural resource management
NSW: New South Wales
OECD: Organisation for Economic Cooperation and Development
PC: Productivity Commission (Australian)
PIMC: Primary Industries Ministerial Council (Australia). This body is made up of the ministers responsible for primary industries in each of the states and territories and the Commonwealth government.
PPD: Presidential policy directive
PRF: Pasture, rangeland, forage (plan of insurance) (US)
QCCA: Queensland Centre for Climate Applications
QCCCE: Queensland Climate Change Centre of Excellence
RASAC: Rural Adjustment Scheme Advisory Council (Australia)
RDCs: Research and development corporations (Australia)
RDEWS: Regional drought early warning information systems
RMA: USDA Risk Management Agency
SECC: Southeast Climate Consortium (US)
SOI: Southern Oscillation Index
UCRB: Upper Colorado River Basin
UNCCD: UN Convention to Combat Desertification
USDA: US Department of Agriculture
USDM: US Drought Monitor, www.droughtmonitor.unl.edu/
VegDRI: Vegetation Drought Response Index, www.vegdri.unl.edu
WA: Western Australia
WATF: Colorado Water Availability Task Force
WFP: United Nations World Food Programme
WGA: Western Governors' Association (US)

1

Science, Policy, and Wicked Problems

Linda Courtenay Botterill and Geoff Cockfield

CONTENTS

This book is a study of drought science and policy and the interface of that science and policy in two developed countries, Australia and the United States. These countries have been chosen because of some general similarities and some key differences in approaches to both the science and policy of drought. Both have well-developed scientific research on climate and droughts. Both have highly variable climatic conditions across large areas that include relatively dry regions, though these are more extensive in Australia. Both have high-technology, extensive and intensive agricultural production with a reliance on irrigation water, and the farm sectors are considered to be both economically and socially important. In terms of policy development, both are federal, democratic systems, each with two dominant political blocs that, to varying degrees, accept the general notion of a mixed economy. Setting aside these factors allows the authors in this volume to focus on the possible reasons for differences in drought science and policy, which will yield insights and lessons in regard to the relationship between scientific advice and the policy process.

In particular, we consider why Australia is the only country with a national drought policy nominally based on the principle that drought is a normal part of climate and a risk to be managed. On the other hand, the United States, with the National Drought Mitigation Center and the National Integrated Drought Information System, undertakes the most comprehensive drought monitoring and research in the world. By contrast, Australia has struggled to establish an authoritative process for reporting on and monitoring drought. Another key difference is that Australian drought policy is largely focussed on agriculture whereas the United States takes a much broader view, encompassing municipal water management and the needs of the tourism industry.

This chapter discusses the key themes that run through the contributions to this collection and we begin the chapter by addressing one of the perennial problems associated with discussing drought: the search for an agreed definition of the phenomenon.

Defining Drought

Droughts are most obviously periods of relatively low rainfall. Hennessy et al. (2008), drawing on the American Meteorological Society statement of 1997, identify four 'types' of drought: meteorological, indicated by low rainfall and high net evaporation; agricultural, where there are low levels of soil moisture, so the production of food and fibre decreases; hydrological, where there is a reduction in surface and subsurface water supply; and socioeconomic, whereby there are adverse effects on 'the supply and demand of goods' and 'human well-being' (Hennessy et al. 2008, 3). These are, however, not discrete types and as Boken (2005, 3) argues, a 'meteorological drought is just an indicator of deficiency in precipitation, whereas hydrological and agricultural droughts are physical manifestations of meteorological drought' and, we would add, a socioeconomic 'drought' follows from these physical manifestations. Nonetheless, these categories are useful in highlighting the range of possible indicators of the extent of impacts, from primary measures of rainfall deficiency (meteorological science) to crop yields (agronomic sciences), dam levels (hydrology), household income (economics and finance), and well-being (psychology and sociology). It should be noted that there are no agreed threshold points for rainfall, water levels, yields, or income that denote when dry conditions become a 'drought'.

This is further complicated if there is no agreement on which of the indicators, and implicitly which type of science, should determine the official recognition or declaration of a drought. There have been efforts to combine indicators in sets of criteria that are used to 'declare' droughts that then trigger government action. For example, during the 1990s, the Australian government's Drought Exceptional Circumstances declarations were based on, among other things, a combination of: meteorological conditions, agronomic and livestock conditions, farm income levels, and the scale of the event (Bedo 1996, 19). By 1999, farm income was the threshold indicator with much more emphasis on the 'exceptional' elements, to the extent that the frequency of the event (one in 20–25 years) was a key consideration (ARMCAN2 1999, 63). Frequency indicators, while deriving from probabilistic data, have some problems in application.

First, looking backward, a particular frequency indicator may not match popular or industry perceptions of drought occurrences. As an example, Figure 1.1 shows the cumulative rainfall deficit/surplus, summing the differences between actual and average rainfalls by month at the Condamine River weather station in the Darling Downs region of Queensland, Australia, that are typical for this agricultural area. The ovals in Figure 1.1 show the periods that have generally been considered droughts. Over 140 years, there have been 11 historically accepted droughts, which is an average of one event every 12.7 years,* twice as frequent as the Australian government's intended threshold. Governments will have difficulty in getting those affected by drought to shift to a more stringent threshold, especially as droughts are also determined by media representations (Cockfield and Easton 2003; Wahlquist 2003), the mobilisation of cultural myths (Botterill 2006), and political lobbying.

The second point is that a particular threshold or guideline is implicitly based on some notion of a normal or acceptable range in which individuals and businesses do not need additional advice, support, or assistance and the official droughts are restricted to the steepest of the rainfall deficiency 'troughs', as in Figure 1.1. If however, rainfall is declining over time, as has been suggested is the case for much of the southern and central agricultural areas of Australia, including the Darling Downs (see Chapter 3), then this creates some definitional and policy dilemmas. If, for example, the Australian 'millennium' drought (2002–2009) (see Figure 1.1) is a harbinger of the future, then, comparatively, there will be longer sequences of dry years and much less soil moisture. This could mean, given no other changes in land and water use practices, lower crop and livestock yields, less water, and lower farm, regional, and national incomes, presuming commodity prices do not increase to offset the decrease in output. If the occurrence criterion stayed at one in 20–25 years, then this is an implicit transfer of the risk management to resources users. On the other hand, there will be increasing pressure on governments to mitigate the effects of droughts according to historical memories of 'normal' conditions.

Considering these two points and the previous description of the different types of droughts, we use a modified approach for this book. A drought is here considered as one aspect of climate variability, denoted by some combination of rainfall, water levels, farm yields, and/or income 'troughs' or low points. Furthermore, we consider that each period of severe moisture deficit within the variation in climatic factors will have hydrological, agronomic, and socioeconomic consequences. We therefore adapt the typology noted above to think in terms of drought science and policy as requiring consideration of meteorology, hydrology, agricultural production, and socioeconomic

---

* The drought records and these data also suggest that approximately 27.8 percent of years are 'drought' years.

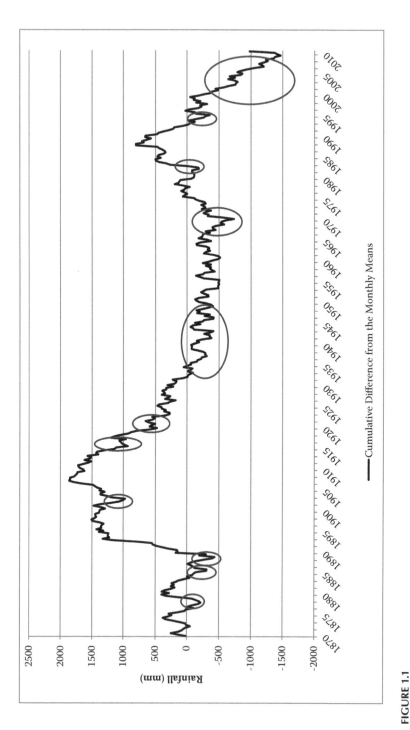

FIGURE 1.1
Rainfall deficiency at the Condamine River weather station (Two of the monthly measurements were from the nearest weather station to replace missing data.) and generally acknowledged droughts.

trends and conditions (see Table 1.1). This breadth goes to a theme in this book that the foci of drought science and policy vary across time and jurisdictions. The science is in developing indicators of drought severity, seeking climatic patterns and trends, forecasting, and research for adaptation to climate variability. The policy can be considered in two broad categories: reactions to particular drought events and strategies to minimise the impacts of droughts in general through mitigation and preparation activities (Table 1.1).

TABLE 1.1

Summary of Scientific and Policy Foci for Droughts

Focus	Indicators	Policy Responses	
		Reactive	Anticipatory
Meteorology	Rainfall deficiency	Drought declaration	Forecasting Climate research
Hydrology	Water availability	Drought declaration Water rationing Moral suasion on water use	Forecasting water availability Water storage construction Water use efficiency research
Agriculture	Soil moisture Production of food and fibre Soil condition/erosion	Drought declaration Farm-level subsidies	Seasonal forecasting Plant breeding for dry conditions Research on resources protection
Society and economy	Household income Psychological well-being Regional/state/national income	Drought declaration Welfare safety nets Regional employment programmes Counselling services	Income equalisation programmes Economic extension services Social research
Human habitats	Fire risk: • Fuel load • Vegetation dryness • Atmospheric dryness	Fire prevention strategies Fire fighting systems	Vegetation management strategies Household protection plans Building regulations and land use rules Fire management research
Other species' habitats	Fire risk: • Fuel load • Vegetation dryness • Atmospheric dryness Habitat condition Food availability Species numbers and diversity	Fire prevention strategies Fire fighting systems Species relocation Supplementary feeding Culling	Vegetation management strategies Corridors Control of other disturbance processes Ecological research

In Table 1.1 we also expand on the conventional categorisations of drought to include consideration of human and nonhuman habitats. In particular, there is the impact of fires on both types of habitat, as well as the threat to nonhuman food supplies. Extended dry periods are strongly associated with major fires, which have both ecological and social impacts, and also with contributing to the decline of habitat condition, especially where other threatening processes, such as landscape fragmentation, are occurring. For this book, we largely stay with the conventional drought types because this is where policy responses have been generally directed, but there is some discussion in Chapter 3 of the habitat effects. Within the conventional categories, there is also an inevitable bias to agricultural considerations because that is where much of the evidence about the science/policy interface can be found. Agriculture has been the dominant focus of Australian policy and remains important in the United States and, since both countries have extensive agricultural research systems, farmers are end users of drought information who have been exhaustively studied.

Nonetheless, in agriculture and water, habitat and fire management, the scientists are engaged with policy makers, providing advice on the validity of indicators, and the effectiveness of policy responses. In effect, scientists are part of 'communities' engaged in policy development. A policy community includes all those with an interest in an issue and a capacity to influence state responses, including pressure groups, government ministers, and experts (Davis et al. 1993, 144). Hence, in a drought policy community, government officials, members of the Executive branch of government, and other experts in relevant government agencies must work with, or at least work with the presence and influence of, the farm lobby, water users, and perhaps even welfare organisations.

Table 1.1 illustrates that there is a range of areas in which scientific (including social scientific) research and expertise has a clear role to play in the development of societal response to climatic variability and uncertainty. However, as the literature on the sociology of science and, more recently, on evidence-based policy making demonstrates, the relationship between science and society is not straightforward.

Science and the Policy Process

The conversation between policy and science is two-way. The majority of the critiques of evidence-based policy making have focussed on the failure of policy makers to utilise effectively the evidence that research provides (for an exception to this approach, see Botterill and Hindmoor 2012). Also important to consideration of the integration of drought science into drought policy is the role scientists perceive for themselves in the policy process. An important contribution to the literature on the role of scientists in the policy process is Pielke's (2007) work. Pielke presents a typology of scientific engagement with the policy process that is useful in understanding the interaction between research and policy. He suggests four possible roles: the *pure scientist*, the *science arbiter*, the *issue advocate*, and the *honest broker of policy alternatives*. These roles vary from providing scientific information and leaving it to policy makers to use it if they wish (pure scientist) to actively pushing a particular solution (advocate) or enabling the question to be better defined for better decision making. Pielke sees a role for all four types of science in the policy process. Importantly, he recognises that policy is often about values and he argues that the more value-laden a policy debate is, the less likely it is that scientists can make a valuable contribution. His colourful contrasting of 'abortion politics'—meaning issues where there is a strong polarisation of values—with 'tornado politics'—whereby communities tend to unite to confront a common threat—provides a helpful illustration of the limits of as well as the potential for scientific engagement in decision making.

Drought policy is played out against a backdrop of a range of potentially conflicting and sometimes deeply held values. A deeply held set of values and beliefs is the often unacknowledged agrarianism in which farming is a highly valued mode of production and way of life and farmers are seen to have desirable personal characteristics such as thrift, industry, self-sufficiency, independence, and wholesomeness (Flinn and Johnson 1974; Aitkin 1985; Botterill 2006; McAllister 2009). Central to the modern version of agrarianism is the importance of the family farm as the preferred form of agricultural organisation and a concomitant sympathy toward policies that keep farmers on the land. For those steeped in agrarianism, there is also a suspicion of expertise that has clear implications for the acceptance of scientific advice in policy decision making, especially where this advice has the double disadvantage of coming from the educated elite and 'city' organisations and thinking.

Agrarianism is often characterised by a perceived cleavage between urban and rural people. This stems from a fundamental view of agriculture as the foundation of society and economy that enabled settlement and consequently civilisation. 'Progress', however, has not necessarily been seen as an advantage for the rural folk, since subsequent industrialisation and urbanisation shifts power and services to cities. The consequent sense of inequity partly informs arguments about the provision of assistance to farmers during drought, even in cases where many farmers are actually wealthy in asset terms.

Competing with the agrarian discourse is the view that farming is an economic activity like any other and, in an era of economic liberalism, used in the British political theory sense, farmers should not expect to be subsidised when faced with risks that can be expected and planned for. A further value, sometimes in competition with agrarian sentiment but similarly running counter to the economic argument, relates to environmental concerns about the impact of drought on the natural resource base and the need for government intervention to ensure it is sustained. All of these values are evident in policy debate and balancing these competing interests is central to the task of the policy maker (Botterill 2004; Thacher and Rein 2004; Stewart 2006). One of the consequences of that balancing act is that not all of the demands on policy makers can be met. This means that 'the science' does not always trump other considerations; an outcome that, from the scientist's perspective, is suboptimal may be the best outcome for decision makers in terms of meeting the needs of all of the values 'watchdogs' (Lindblom 1959) in the policy debate.

Young et al. (2002) propose five models of the way in which evidence is used in policy: the knowledge-driven model, the problem-solving model, the interactive model, the political/tactical model, and the enlightenment model. These five types can be useful in examining the US and Australian experiences with drought science and policy. Young et al. (2002, 216) describe the knowledge-driven model as an approach that 'assumes that research *leads* policy'. To a large extent, this describes the situation in the United States. Drought researchers, particularly at the National Drought Mitigation Center, have led the policy debate, providing research and communicating their views about the need for a national drought policy to be implemented. The second type the authors propose is the problem-solving model. In this model, 'research *follows* policy'. This type fits neatly the development and implementation of the Australian National Drought Policy. As will be explained in Chapter 6, Australian policy makers developed the exceptional circumstances programme and then turned to scientists to assist in the development of criteria for deciding when such circumstances existed.

Young et al. (2002) note the limitations of these two, essentially linear models of the policy process and our cases appear to bear this out. In the US case, the scientists advocating for a risk-management-based drought policy have lacked a political champion who could have ensured that the policy process responded to the science and resulted in policy change. In the Australian case, the exceptional circumstances declaration criteria never achieved sufficient legitimacy to avoid debate and conflict over the declaration process.

The third model in the typology is the interactive one, which Young et al. (2002, 217) argue posits 'a much more subtle and complex series of relationships between decision makers and researchers'. This model sees the relationship as 'mutually influential' involving a policy community of actors with a variety of roles. It is arguable that the National Integrated Drought Information System set up in the United States in 2006 fits this description as it draws on a wider community of actors with an interest in drought monitoring than simply the scientific research community.

Young and colleagues' fourth model is the political tactical model, which is similar to Pielke's issue advocacy idea in that the research agenda is 'politically driven'; in the Young et al. model, however, the political agenda is explicitly that of the government of the day, whereas Pielke's issue advocate can be employed by nongovernmental actors to promote a particular political agenda. The final model (preferred by Young and his colleagues) is the enlightenment model, which shares similarities with Pielke's honest broker. The enlightenment model 'portrays research as standing if not aloof, then certainly a little distant from the hothouse of immediate policy concerns. Rather than research serving policy agendas in a direct fashion, the benefits are indirect' (Young et al. 2002, 217).

Although many researchers may prefer to stand at a distance from the messiness of the policy process, the interactive model might be more appropriate in a case such as drought where the science is clearly so important. An interactive approach would see the science engage stakeholders, as in the NIDIS model, and therefore win their endorsement of its authority. This would then be integrated into a policy response that also recognises the authority of the agreed science. In the Australian case, this model holds promise as it suggests that a trigger that has sufficient credibility and buy-in from stakeholders can take much of the political heat out of the declaration process. For this model to work, scientists may need to accept a greater role in the policy process, to some extent entering an area that Weinberg (1972) described as 'trans-science'. Weinberg's identification of trans-science recognised that values can and do intrude into decision making and that although trans-scientific issues are

> epistemologically speaking, questions of fact and can be stated in the language of science, they are unanswerable by science; they transcend science. In so far as public policy involved trans-scientific rather than scientific issues, the role of the scientist in contributing to the promulgation of such policy must be different from his role when the issues can be unambiguously answered by science. (Weinberg 1972, 209)

Given the broader considerations of the impacts of drought and its social and environmental consequences, determining the severity of drought and the appropriate societal response requires judgements that go beyond the purely 'scientific' and include values issues.

Drought as a Wicked Problem

A further framework for considering the complexity of drought as a policy problem is the concept of the 'wicked problem', first articulated by Rittel and Webber (1973). The 'wicked' aspects of drought are the following:

- There is no definitive formulation of the problem.
- The problems have no stopping rule.
- Solutions are not true or false, but rather bad or good.

- Every solution is a "one-shot" operation; because there is no opportunity to learn by trial and error, every attempt counts significantly.

Fundamentally, drought becomes a problem when it impacts on people's quality of life. However, in some cases, drought can exacerbate underlying problems relating to the economics of individual farm operations, such as subviable size or poor natural resource management problems. The 1992 approach to Australian drought policy sought to address this aspect of the problem by offering support during drought only to those farmers with long-term prospects in agriculture. This recognised the essential 'wickedness' of the problem in that the magnitude of the impact of drought may in fact be a symptom of, and interacting with, another problem, such as poor farm management. Within 2 years there was a change of direction with the addition of a welfare component in 1994 that was not tied to farm performance. Reflecting the idea that solutions for wicked problems are bad-or-good, the public reaction to the early years of the drought indicated that the policy was certainly considered inadequate, if not outright bad. Frequent incremental adjustments to the definition of exceptional circumstances support the application of Rittel and Webber's term to the Australian situation. Brian Head sums up Rittel and Webber's argument: Wicked problems are "inherently resistant to a clear statement of the problem and resistant to a clear and agreed solution." It could be argued that the fundamental problem, reduced rainfall, is clearly understood, but as will be argued in Chapter 2, exposure to drought and other risks is heightened or lessened by social and economic choices. Science cannot resolve these dilemmas by filling the gaps in empirical knowledge. (Head 2008, 102). In this sense, wicked problems are trans-scientific.

A 2007 paper by the Australian Public Service Commission (APSC) on wicked problems drew attention to the need to work across organisational boundaries in addressing them (APSC 2007, 35–36). To date some of the more difficult elements of drought policy in Australia (for example, the relationship between drought and water allocations and between drought and climate change) have been subject to institutional 'firewalling' (Thacher and Rein 2004). As noted before, drought policy in Australia has been predominantly agricultural policy; it has focussed on farmers and farm management. The policy framework within which the National Drought Policy was developed was one of facilitating structural adjustment in agriculture and the component policies of the drought approach were targeted at farmers. The focus on agricultural producers has limited the policy community engaged in drought policy debates; it has not included urban water managers or other sectors of the economy, with only belated recognition of the impact of drought on farm-dependent small businesses in rural communities.

While this small policy community was able to reach agreement in the early 1990s on the drought policy approach, it was limited in its capacity to deal appropriately with the welfare dimensions of drought and it did not engage with debates around water management policy or climate change. Debates around the management of the Murray–Darling Basin underway at the time of writing continue to take place independently of current reviews of the National Drought Policy. By contrast, the US drought policy debate has been more encompassing, with the impact on agricultural producers comprising only part of the discussion. Thacher and Rein's work on firewalling would suggest that, while it would appear sensible to take a holistic approach to a multidimensional complex issue, it limits the capacity to achieve agreement on appropriate policy because of the large range of potentially conflicting interests, stakeholders, and values that are involved.

Communication at the Interface

The relationship between scientific research and the policy process is further complicated by the difficulties of communicating between the two worlds. This section identifies a number of key issues that arise in communication at the interface between science and policy.

The first is a simple problem of language. Terms used in policy documents that are not precise or have multiple meanings can obfuscate and confuse policy debate. Scientists and policy makers apparently speaking the same language can be talking at cross-purposes, and the communication of science necessarily involves a simplification of the research in order to make information digestible and to provide policy makers with evidence in a form that is usable. Ludwik Fleck explained the need to summarise science as follows:

The words that are used in policy and science communication can also confuse rather than enlighten. Two examples from recent drought policy debate in Australia are 'risk' and 'resilience'. Australia's drought policy approach is based on a risk management approach but the term 'risk' is not uncontested. Although its original meaning was associated with probability and gambling—and therefore associated with the idea that a risk could be managed to the benefit of the manager—recent work, notably that of Ulrich Beck and his concept of the 'risk society' (Beck 1992), has imparted to the term a much more negative set of associations. As Botterill and Mazur note (2004, 15),

> While risk might be a term used widely across society, its meaning varies and remains somewhat contested. Over the last two decades, research has found that 'risk' is often now associated with 'bad risk'. The shift away from the original technical probabilistic meanings of the term has significant implications for risk assessment, management and communication processes.

Hayman and Cox (2003, 156–157) likewise note that 'economists, scientists and farmers struggle to find a common language for concepts of risk management, especially on an issue as emotive as drought' and the word 'risk' is part of the problem as it can be used in 'both a technical-legal sense and in everyday language'.

A more recent addition to the Australian drought policy vocabulary is the term 'resilience'. With its origins in mathematics, the term has crossed disciplinary boundaries into a variety of fields and its meaning has changed in important ways as this has occurred. As Klein et al. (2003, 40) note, 'What was once a straightforward concept used only in mechanics is now a complex multi-interpretable concept with contested definitions and relevance'. Resilience in social systems means an ability to recover and function well after systemic disruptions. Building resilience is about developing people's capacity to adapt but such an approach can also signal a change in government attitude to risk, in that there is a shift from the safety net to enabling self-sufficiency.

A second problem arises in that the status of scientists as authoritative sources of 'the truth' has diminished over the past century. As Yearley (2005, 111) notes, recent decades 'have witnessed repeated conflicts over the public's trust in and acceptance of scientific expertise'. Scandals such as the apparently slow official response to 'mad cow disease' (bovine spongiform encephalopathy [BSE]) have raised doubts in the public's minds about scientists' capacity to consider all the relevant factors in providing advice, particularly in relation to risk. Jasanoff (2003, 160) writes of the experts' tendency to demarcate or frame problems so that they are 'tractable to analysis'. A consequence of this framing is that 'what lies within the perimeter of expert competence tends to be labelled 'science' or 'objective' knowledge; what lies outside is variously designated as values, policy or politics'. This is compounded by the recent problems science faced with respect to its credibility (Yearley 2005). The uncertainties inherent in science and the capacity of the legal system to exploit this in the courtroom have contributed to a perception of science as contested, rather than definitive. Adversarial legal systems are, as Yearley (2005, 144) describes them, 'peculiarly suited for the deconstruction of scientific expertise, even of expertise regarded as reputable within the scientific community'. All of these factors point to the need to consider both science and politics in policy making in all but the narrowest of policy arenas; reinforcing the argument that the interactive model of science–policy relations may be the best way forward.

For drought policy, a general problem with the status of science more broadly can be compounded by the agrarian mistrust of expertise. In their seminal work on agrarianism, Flinn and Johnson (1974, 194) argue:

> Within agrarian belief there is pride, a certain nobility, in what man accomplishes by the sweat of his brow. There is suspicion about a man who makes a living using his head and not his hands. There is a feeling that everyone could and should graduate from high school, but colleges—too much education is not good.

There is something of a contradiction here in that farmers are quick adopters of technologies that increase their productivity and are responsive to farm problems and, as will be discussed in Chapter 6, early government investment in science in Australia was focussed on agriculture. However, Australian farmers also responded very negatively to the scientific advice that was presented to them about the need for reduced water consumption in the Murray–Darling Basin and those at public meetings across the Basin were quick to argue against the science and, at times, vilify the scientists.

Drought provides an interesting case study of the challenges at the science–policy interface. The following discussions are divided into three parts: *Managing Risk; Science, Evidence, and Policy;* and *From Policy to Action*. The authors have a common normative position that drought is a normal part of climate and a risk to be managed. This is also reflected in policy terms in the Australian National Drought Policy and the US National Drought Policy Commission's Report. Section I of the book considers this element of drought management—that is, as a risk to be managed, including the implications of the predicted impacts of future climate change. The second part considers the policy response to these challenges, as well as giving further attention to the role of scientific input into the policy process. Section III considers drought risk management in action to look at the take up of drought science by end users in the community. We conclude with some final observations about the lessons that can be drawn from our drought case studies about science, policy, and managing uncertainty.

References

Aitkin, D. 1985. 'Countrymindedness'—The spread of an idea. *Australian Cultural History* 4:34–41.

APSC. 2007. Tackling wicked problems: A public policy perspective. Canberra: Commonwealth of Australia.

ARMCAN2. 1999. Record and resolutions: Fifteenth Meeting Adelaide 5 March 1999 Canberra: Commonwealth of Australia.

Australian government. 2011. Exceptional circumstances, ed. Department of Agriculture Fisheries and Forestry. Canberra: Australian Government.

Beck, U. 1992. *Risk society: Towards a new modernity. Theory, culture & society*. London: Sage.

Bedo, D. 1996. Rainfall decile analysis and drought exceptional circumstances. Paper read at Indicators of Drought Exceptional Circumstances, October 1, at Canberra.

Boken, V. K. 2005. Agricultural drought and its monitoring and prediction: Some concepts. In *Monitoring and predicting agricultural drought: A global study*, ed. V. K. Boken, A. P. Cracknell, and R. L. Heathcote. Cary, NC: Oxford University Press.

Botterill, L. C. 2004. Valuing agriculture: Balancing competing objectives in the policy process. *Journal of Public Policy* 24 (2): 199–218.

————. 2006. Soap operas, cenotaphs and sacred cows: Countrymindedness and rural policy debate in Australia. *Public Policy* 1:23–36.

Botterill, L., and N. Mazur. 2004. Risk and risk perception: A literature review: A report for the rural industries research and development corporation. Canberra: RIRDC Publication No 04/043.

Botterill, L. C., and A. Hindmoor. 2012. Turtles all the way down: Bounded rationality in an evidence-based age. *Policy Studies*. doi: 10.1080/01442872.2011.626315.

Cockfield, G., and K. Easton. 2003. Reporting the drought: A study of news cycles. *Australian Journalism Review* 25 (1): 171–186.

Davis, G., J. Wanna, J. Warhurst, and P. Weller. 1993. *Public policy in Australia,* 2nd ed. St Leonards, NSW: Allen & Unwin.

Fleck, L. 1979 [1935]. *Genesis and development of a scientific fact.* Translated by T. J. Trenn and R. K. Merton. Chicago: University of Chicago Press.

Flinn, W. L., and D. E. Johnson. 1974. Agrarianism among Wisconsin farmers. *Rural Sociology* 39 (2): 187–204.

Hayman, P., and P. Cox. 2003. Perceptions of drought risk: The farmers, the scientist and the policy economist. In *Beyond drought: People, policy and perspectives,* ed. L. C. Botterill and M. Fisher. Melbourne: CSIRO Publishing.

Head, B. 2008. Wicked problems in public policy. *Public Policy* 3 (2): 101–118.

Hennessy, K., R. Fawcett, D. Kirono, F. Mpelasoka, D. Jones, J. Bathols, P. Whetton, M. Stafford Smith, M. Howden, C. Mitchell, and N. Plummer. 2008. An assessment of the impact of climate change on the nature and frequency of exceptional climatic events. Canberra: Bureau of Meteorology and CSIRO.

Jasanoff, S. 2003. (No?) Accounting for expertise. *Science and Public Policy* 30 (3): 157–162.

Klein, R. J. T., R. J. Nicholls, and F. Thomalla. 2003. Resilience to natural hazards: How useful is this concept? *Environmental Hazards* 5:35–45.

Lindblom, C. E. 1959. The science of 'muddling through'. *Public Administration Review* 19:79–88.

McAllister, I. 2009. Public opinion towards rural and regional Australia: Results from the ANU Poll. *Report 6,* October 2009.

Pielke, R. A., Jr. 2007. *The honest broker: Making sense of science in policy and politics.* Cambridge: Cambridge University Press.

Rittel, H. W. J., and M. M. Webber. 1973. Dilemmas in a general theory of planning. *Policy Sciences* 4:155–169.

Stewart, J. 2006. Value conflict and policy change. *Review of Policy Research* 23 (1): 183–195.

Thacher, D., and M. Rein. 2004. Managing value conflict in public policy. *Governance* 17 (4): 457–486.

Wahlquist, A. 2003. Media representations and public perceptions of drought. In *Beyond drought: People, policy and perspectives,* ed. L. Courtenay Botterill and M. Fisher. Collingwood: CSIRO Publishing.

Weinberg, A. M. 1972. Science and trans-science. *Minerva* 10:209–222.

Yearley, S. 2005. *Making sense of science.* London: Sage.

Young, K., D. Ashby, A. Boaz, and L. Grayson. 2002. Social science and the evidence-based policy movement. *Social Policy and Society* 1 (3): 215–224.

Section I

Managing Risk

2

Risk, Expertise, and Drought Management

Geoff Cockfield

CONTENTS

The drought policies and scientific modelling and extension discussed in this book are instruments of risk management, provided by governments and research agencies to influence the behaviour of individuals, businesses, and households. Some science—for example, that used to quantify the risk of dry periods—is directed at resources managers so that they develop systems, such as minimum tillage farming, dams of an appropriate size, and alternative income for agricultural service businesses to cope with variability over time. Other scientific work focuses on analysing and monitoring natural phenomena, such as the El Niño Southern Oscillation (ENSO) System, in an effort to provide early warnings so that managers can, for example, institute water restrictions, sell off livestock, or cull dominant species in a national park. Governments, informed by the economic and other social sciences as to the scope and scale of effects, try to influence the behaviour of the resources managers and those affected to ameliorate the impacts of the drought. Drought is normally classified as a 'natural' hazard, with the implication that little can be done to prevent each event. In this chapter, I argue for three qualifications of that narrow definition. People's life and business choices influence the degree to which the base physical phenomenon, little or no rainfall, is a hazard to them. Second and related to that, droughts are also culturally and politically defined. Finally, if climate change contributes to an increase in the frequency and severity of droughts, then they are at least partly a technological risk; that is, there is some causation related to human agency.

The chapter starts with some definitions of risk and the three qualifications to considering drought as a natural hazard are further elaborated. The reasons governments become involved in risk management through the provision of research, extension, and policy are critically considered. Then approaches to estimating and communicating risks are discussed. In general, while there has been a move away from technocratic dominance to greater involvement of lay people, I note some arguments for the pendulum to swing away from excessive participation in risk management in certain circumstances. Complicating this, though, is the rise of rationalist approaches, in which the economic costs and benefits of risk management are considered before any action is taken. This is evident in attempts to shift the risks of drought from government to nongovernment sectors, as is discussed in several chapters of this book.

There are, however, still countervailing forces for government intervention, notably claims that the agricultural sector has special economic and social roles, as discussed in Chapter 1. The examples in the first part of this chapter are primarily drawn from research into agricultural drought, since this has been a particular policy focus, especially in Australia. Some of the authors in this volume consider impacts on other sectors and the chapter concludes with the contention that a broader view of drought sciences and policy opens up a different set of arguments about government involvement in drought risk management. In particular, multisector drought policies and science involve collective goods, not just the private goods of farms and agriculture-dependent businesses.

Drought as a Risk

According to Fischhoff and Kadvany (2011, 5), 'Risks threaten things we value. What we do about them depends on the options we have...the outcomes we value...and our beliefs about the outcomes that might follow, if we choose each option'. The degree of risk is determined by a combination of the probability of occurrence and the severity of the impacts (Adams 1995, 8). Risk has been narrowly defined to include events where the probabilities are known with all else considered as uncertainty, although Adams (1995, 25-27), argues that uncertainty should also be included in a broad definition of risk. Risks are often categorised according to origins, with natural hazards being events where there is 'a fluctuation in the balance between mankind and nature to the disadvantage of mankind' (Lofstedt 2005, 86), and such events are seen as unpredictable and uncontrollable (Axelrod et al. 1999, 32). This can be a factor in how risks are perceived since risk tolerance is higher if the event or threat is seen as being natural, familiar, and voluntarily undertaken (Slovic 1987, 281; Renn 2008, 109). Drought is familiar and natural to most resource users and managers, though it may not necessarily be considered voluntary if those affected see themselves as being culturally and economically constrained in their life choices.

Drought seems to fit the definition of a natural hazard, since rainfall cannot yet be controlled by humans and there is significant disadvantage to people, while the risk can be quantified through estimating the probabilities of frequency and severity of droughts from meteorological records. There are, however, some qualifications to this notion of 'natural'. First, droughts are at least partly socially and politically defined, as is argued in various chapters of this volume. That is, there are particular periods of time with little or no rainfall that are identified as drought by farm organisations, resources managers, and journalists and then there are government declarations for administrative purposes, such as triggering programmes (see Chapters 4, 6, 10, and 11), and the official definitions can change over time. Hence, and following Renn (2008, 3), drought like other risks has both objective (meteorological measurements) and subjective elements.

The second qualification to the notion of naturalness is that the impacts of drought, or any hazard, can be moderated by local expectations and actions (Adams 1995, 30); for example, people may live nearer to or further from a flood-prone area. A famous Australian example of this in relation to drought is Goyder's line, which was surveyed in 1865 as the northernmost 'boundary' for grain production in South Australia. A subsequent run of good seasons resulted in increased crop production and settlement north of that line, but later drier seasons resulted in a return to extensive grazing, leaving behind abandoned villages and farmhouses. The probability of an agricultural (crop) drought increased because of human choices in regard to marginal land uses. George Goyder, the South Australian surveyor general, did the risk analysis through a land survey, which was initially ignored, but the analysis stood the test of time. Adjustments follow the mounting evidence since people come to accept risks and then they are more able to deal with them (Renn 2008, 107).

The third qualification to the natural hazard categorisation is that climate change may influence the frequency and severity of drought (see Chapter 3 for more on this) and the impacts of climate change are technological, rather than natural risks. Technological risk analysis and management, which dominate the risk literature, tend to focus on case studies involving pollution, side effects from pharmaceuticals, nuclear power, transport, and traffic and so on (see, for example, Renn 2008; Sjoberg 1998; Slovic 1998; Wildavsky and Dake 1998). The increasing concern about technological, or manufactured, risks is such that some researchers have identified that we now live in a 'risk society' (Beck 1999). A risk society is 'increasingly pre-occupied with the future (and also with safety)' (Giddens 1999, 3) and since modern industrial societies operate on a 'high technological frontier' (Giddens 1999, 3), there is increasing recognition of and concern about 'manufactured risk' (Giddens 1999, 5), or the risk that results from human actions and decisions (Beck 2009, 293). Hence, there is a demand for governments to ameliorate these risks.

Climate change is a rather unusual technological risk due to its scale, and it can really only be mitigated with collective global action, so policy coordination is extremely difficult for national governments. Hence, even though there is the potential for amelioration, there can be an understandable feeling of helplessness, just as there is with floods or earthquakes, because of the scale of the problem and because it is an extension of a natural phenomenon, the greenhouse effect. Because of these factors and the uncertainty around actual impacts on climatic events, the case to redefine drought as even partly a technological risk is as yet weak. The authors in this volume say no more than that policy makers will need to factor the science of climate change into drought research, and perhaps policy, but there may be potential for climate change mitigation to also mitigate drought, which goes to the issue of the breadth and boundaries of drought sciences and policy.

In summary, drought is generally considered a natural hazard and despite the proposed qualifications in this chapter, will likely continue to be seen that way, though both the Australian and US governments want more emphasis on the 'natural' than the hazard (see Chapters 4–7 and 10–12). The definition of drought is extremely important to those in the relevant scientific and policy communities, since that definition can determine government responses and transfers to affected parties, and data and modelling requirements (see Chapters 4–7, 10, and 11). An important side effect of the risk society, though, is the questioning of science, with climate change debates being an example of quite heated questioning (see Chapter 3). All this makes for complex science/policy/end user interactions and it is understandable that governments might be reluctant to get involved in drought risk management. This is especially the case with governments, such as those in the United States and Australia, who are also seeking to limit taxation and therefore expenditure, in the interests of economic growth and efficiency, not to say ideology.

The Roles of Governments in Risk Mitigation

An important element of attempts to shift from reactive to proactive drought policies, as described in numerous chapters in this volume, is the shift in political economy in the developed world, and especially the Anglo-American sphere. The rise of market liberalism (known as neo-liberalism in Australia) emphasising economic efficiency, self-reliance, and individual liberty is well known and drought policy has not been immune from that trend. There was, however, something of a recent countertrend against the generally increasing involvement of governments of developed countries in risk management throughout the twentieth century. Using the case of the United States, Moss (2002) argues that governments were increasingly drawn into risk management from the mid-nineteenth century. In the late 1800s, policies, such as limited liability for companies, were enacted to redistribute risk and encourage investment (Moss 2002, 4–5). Then there were additional protections for workers (1900 to 1960) and finally 'security for all' through an expanding welfare system (Moss 2002, 5–6). As will be shown in this volume and in a somewhat similar vein, drought policies in Australia were developed to help preserve the farm business and then later a welfare component was also included (see Chapter 6).

Moss reviews the arguments (2002, 11–45) for government risk management because of (using the language of neo-classical economics) problems in the efficient operation of markets, including misperception of risks and lack of information about those risks. In relation to misperceptions, he notes that people can rely on available information too much or are affected by unwarranted, at least according to mathematical probabilities, optimism (Moss 2002, 43–44). Hence, one of the primary roles of government in relation to drought is in the provision and distribution of information. It is argued in several chapters in this volume that the development and distribution of more information on climatic trends and events and drought management strategies will help end users make better decisions.

While this is a logical response, these potential beneficiaries of the science may be resistant to the data-driven advice for a range of reasons (reviewed in Renn 2008, 99–105). First, there is the 'availability' barrier, whereby events that come immediately to mind are given greater weight, such as the breaking of a drought leading to reduced concern about future droughts. Second, there might be a representation barrier, whereby a singular event such as a major drought determines how the phenomenon as a whole is seen.

Third, there is the desire to avoid cognitive dissonance, which means ignoring or downplaying probabilities that challenge belief systems, such as climate change, as an existential threat to the way of life of farming. One might argue that a life highly exposed to natural forces and dependent on natural resources, such as farming, requires optimism and even perhaps wilful ignorance in order to persevere. On the other hand, for some people, where there is a strong emotional attachment to the activity, there is generally greater risk acceptance (Renn 2008, 109). If these psychological barriers are considered in policy, some risk management may seem rather paternalistic in that governments are accepting that some people will not use available information to ameliorate risk and therefore it is acceptable to develop policy for 'their own good'. This apparent paternalism (Clarke 2002; Gostin and Gostin 2009)—or more frequently now maternalism by way of the 'nanny state'—is particularly criticised by libertarians and market liberals (Jochelson 2006; Calman 2008).

The libertarian arguments are that any forced redistribution of risk, such as taxation revenue used for drought assistance, infringes individual rights, while creating the expectation that governments will come to the rescue, which is a moral hazard that potentially encourages further risky behaviour (Fischhoff and Kadvany 2011, 14). The argument would be that if farmers are reasonably sure that a government will help them out during a drought, then there is less need for preparatory management actions. In a more specific example, the provision of stock feed subsidies (see Chapter 12) might even encourage landholders to increase their stocking rates, which would indeed be a perverse incentive due *inter alia* to the adverse environmental consequences. The parallel liberal economic arguments are that morally hazardous policies help maintain sectoral inefficiencies that would otherwise disappear over time under competitive pressures, while the cost of programmes adds to the burden on taxpayers and limits the potential for growth in the national economy.

In a variation of this political economy perspective some researchers (Adams 1995; Renn 2008) draw on cultural theory to identify different risk responses. Adams (1995, 57–58) identifies four types of respondents:

a. Egalitarians, who strive to keep reducing risk toward zero, because nature is ephemeral and easily disturbed to generate threats (see also Renn 2008, 37)

b. Fatalists, who think there is not much to be done to reduce risks, which are either predetermined or so random as to be unpredictable, other than avoid them when evident (classified as randomised or stratified individuals by Renn 2008, 37)

c. Hierarchists, who assume that people are taking more risks than necessary and therefore authority figures should seek to reduce these (the 'bureaucrats'; Renn 2008, 37)

d. Individualists, who seek to reduce their own risk, but recognise the benefits of risk-taking and are resistant to interventions (the 'entrepreneurs'; Renn 2008, 37)

These categorisations can be used to understand particular individuals or the orientation of policy content or rhetoric. The egalitarians in government would want to spread risk, through taxation and redistribution or social insurance. The hierarchists are the maternalists; the fatalists are more conservative in thinking intervention is largely a waste of effort, while the individualists, equating to liberals, are actively resistant to redistributing risk. These categorisations need to be treated with some care, since their empirical basis is highly contested (Sjoberg 1998), but they serve to illustrate different tendencies that flow on to policy values.

There are, however, arguments for government involvement in risk management that involve consideration of more than individual business preservation and household welfare. First, there is an efficiency argument in that governments have some significant advantages as risk managers: They can spread risk more broadly, which is effectively what happens when tax revenue is used for drought relief programmes; they can punish cheating of risk management programmes, through their legal standing; and they have considerable financial power and flexibility (Moss 2002, 49–50). At a broad systemic level, Moss (2002, 3) argues that governments became involved in risk management in order to help ensure the continuation of capitalism, notably supporting investor confidence and capital security. In relation to agriculture, safety nets for periods in which there is a rapid downturn in incomes or escalation in costs help preserve the farm in the longer run, which maintains both investment and skills.

The market liberal response to that might be that periodic droughts merely accelerate farm aggregations, driven by competitive pressures and innovation, which would see the most efficient and skilful farmers survive anyway. Then again, such adjustment processes also have flow-on effects in terms of regional investment, and regional development has long been a concern of both Australian and US governments.

Regional development in both countries is about more than just self-sufficiency in food production and decentralisation, though these were important considerations in early land use and settlement policies. There are strong cultural elements to support for agriculture and this has been evident in drought policies, as will be shown in later chapters. To set up those arguments, though, I return to the notion of risk as being culturally and politically defined. According to Renn (2008, 4), the tolerability of risk is driven by 'public values, perceptions and social concerns', meaning that there is selectivity in what we should be concerned about and who should receive support.

As will be argued in Chapter 4, Australian drought policy has been shaped by cultural conceptions of the special roles of farmers in society and economy, which were most notably reviewed by Aitkin (1985, 38). Similar narratives and values, generally known as agrarianism, are evident in other countries (Montmarquet 1989) and have been particularly studied in the United States (Flinn and Johnson 1974; Buttel and Flinn 1975, 1977; Willits et al. 1990; Dalecki and Coughenour 1992) and this way of thinking has been linked to special policy treatment for agricultural industries (Skogstad 1998; Botterill 2009). These cultural and political stories of the importance of agriculture feed in to what Renn (2008, 38) describes as the 'social amplification' of risk. That is, a risk such as drought is amplified through media and political rhetoric as hitting the economic and cultural heart of a country, as observed by Botterill (2006, 2009).

Agrarianism, which is antithetical to market liberalism in that it supports the privileging of one sector over another and social and cultural over economic considerations, is one reason why governments continue to be involved in drought policy, especially reactive policy making. Despite efforts to increase landholders' responsibility for risk management through changes to drought policies in Australia and through community engagement and raising awareness in the United States, there is a lot of evidence in this book that the rhetoric of preparedness and self-reliance is not necessarily matched by policy and practice, with governments still using reactive and preferential measures and farmers still expecting that (see Chapter 12).

There are, however, arguments other than those from agrarian narratives and some even with a basis in economic theory, supporting governments remaining involved in drought risk management. In particular, taking a broader view as in this volume, drought affects some important collective goods and its threats go beyond individual farm businesses and households. The collective goods include water supply and other ecosystem services, while the collective threats include bushfires (wildfires). Strictly speaking, most ecosystem services are not public goods, in that they can be privatised, but they are widely considered to be under the stewardship of governments. Hence, a broader definition of drought requires consideration of a broader set of arguments about the possible roles of government, as discussed in Chapter 3.

Approaches to Risk Assessment and Communication

In this final section, I consider the communication of risk at the science/policy/end user interface, using four 'models' of engagement, recognising that these are descriptive points along a spectrum. At one extreme, in the *technocratic* model the scientists provide the information, which drives policy, and then the risks are communicated (Renn 2008, 10). In the *decisionistic* model, science is used for the risk assessment and then technical and socioeconomic considerations are used for a risk evaluation and management, and thence to policy decisions and outcomes; in this model there is an increasing separation between risk assessment and risk management and a potentially greater separation between the experts and the policy makers. With the increasing influence of the social sciences the idea of risk management being a function of both expert analysis and public perceptions became more influential (Renn 2008, 58). The *deliberative* model therefore starts with socioeconomic and political considerations framing the policy and then there is risk assessment and evaluation, with broader policy community participation in the development of risk management and policy outcomes and communication. Fischhoff (1998) identified stages in the development of risk communication. The first stage was just 'getting the numbers right', followed by explaining the data, identifying the benefits, and ultimately including people in the research and analysis, equating with the deliberative model. In the *rationalistic* model, the specialist risk assessors sit alongside the economists, who assess the costs and benefits of risk assessment and avoidance to determine the policy action. That is, the economists may conclude that the risk level is so small that no resources are allocated to assessment, let alone developing policy.

There are a number of arguments for operating more toward the deliberative end of the spectrum. Participation is seen as a democratic value in itself (Lofstedt 2005, 18), or for the more pragmatic, it is a means of gaining social acceptance of the scientific information and institutions (Fischhoff 1998, 136).

Citizen science, enabling people to engage in the collection and even processing of data—for example, through volunteer weather stations—can have an educative function and also cuts the public cost of information gathering. The interface of community and experts can, however, be difficult because of different approaches to and perceptions of risk. First, there are the psychological barriers to receiving information, summarised before, and then there are the professional and language barriers, briefly touched on in Chapter 1. Technical analysts often favour quantitative methods, such as probabilities (Fischoff 1998, 140) while many nonexperts rely more 'on intuitive risk judgements' (Slovic 1998, 31). Probabilities, especially in relation the weather, can be difficult to understand, with many in the public equating high probability with prediction of an event, rather than just likelihood. Lay people may also be more concerned with the 'equity of risk distribution', rather than just the probability and severity of events (Shrader-Frechette 1998, 47). This is not to imply that risk professionals are entirely objective, since they have some of the same biases as others (Shrader-Frechette 1998, 49) and are also influenced, though perhaps sometimes unconsciously, by ideology (Wildavsky and Dake 1998, 111); rather, the professionals use methodologies that minimise, in their minds, these particular barriers.

Response to risk also varies according to the sociodemographic characteristics of the individual or group (Po et al. 2003, 14; Ekberg 2007, 351), though the introductory nature of this chapter restricts the discussion to broad generalisations. Information receivers consider risks through a series of filters, including cultural background, emotional responses and personal beliefs, the heuristics of information processing, and sociopolitical institutions, including trust in those institutions (Renn 2008, 141). The possibility of drought is highly motivational in gaining attention of resources managers but the receiver of information still has to assess the credibility of the information and this depends on personal experience and perceptions of the source (Renn 2008, 100). In the 'risk society' there is a dependence on expert systems but also scepticism about those systems (Ekberg 2007). This scepticism can be amplified by the cultural context and, as noted in Chapter 1, elements of agrarianism promote a suspicion of experts who are seen not to have practical experience (Flinn and Johnson 1974). The increasing attention to climate change will also confound drought risk communication, since scepticism about climate change may be undermining the credibility of meteorological science and scientists (see Chapters 3 and 12).

In addition, because of the inherent uncertainties in science, more research does not necessarily bring a clear resolution but more likely, greater uncertainty (Beck et al. 2003, 21). Once people form a view of risk, these views can be highly resistant to evidence and persuasion, with the perceptions of nonexperts of the degree of risk tending to be quite different from those of technical experts (Slovic 1987). Processes of deliberation may lead to more divergent views as 'evidence' is examined in detail and this further contributes to declining trust (Lofstedt 2005, 21). According to Lofstedt (2005, 4):

> Public trust in policy-makers, industry officials and opinion-shapers
> is declining in western societies....This...appears to be related to...
> social alienation; a lack of social capital; higher levels of education and
> greater availability of information resulting in a more sceptical public;
> increased scientific pluralism leading to confusing messages;...and a
> hyper-critical media.

Interest groups can fuel public distrust of state actors (Lofstedt 2005, 12) but so too can the actions of the policy makers, with reactive responses tending to reduce trust (Lofstedt 2005, 12). Trust is not then necessarily built by providing more information and engaging more with the public, since the former can be minutely examined for perceived inconsistencies, while the latter provides forums to challenge scientists and policy makers (see Chapter 7 for further discussion of this). How then to communicate risks in a 'post-trust society' (Lofstedt 2005, xv)? Surprisingly, perhaps, some argue for moving back up the spectrum toward the technocratic approaches (Lofstedt 2005, 10; Renn 2008, 59–60).

Reasons for this include the difficulty in managing divergent opinions in the community and concern about the potential for participative processes to affect economic investments. Lofstedt (2005, 10) argues that if there is trust in a particular institution of experts, then a more technocratic approach is possible. Trust in an agency cannot be assumed simply because of its longevity or peer rating (Lofstedt 2005, 128) and there still needs to be stakeholder agreement on courses of action (Lofstedt 2005, 129). Nor does rebuilding trust in technocratic institutions mean that stakeholders should be excluded; they can be important in ensuring accountability and that the experts communicate effectively (Lofstedt 2005, 23). There also needs to be trust in the triangular relationship between end users, the experts, and the policy makers. As will be shown in Chapter 7, there can be tension when the experts feel the end users have appealed directly to the politicians, who themselves may distrust the experts (Lofstedt 2005, 25), but this is an inevitable part of democracy.

Conclusions

Drought is and will continue to be considered a natural hazard, though there are some qualifications to that, with the defining of particular droughts being an important point of discussion in this book. As part of democratisation and increasing wealth and perceived value of human security, governments have been increasingly drawn into risk management, though there has been some selectivity in who is protected and the extent to which they are protected. That selectivity is influenced by cultural and political perceptions of which groups are important. Technical analysis is at the base of formal risk assessment but increasingly social science analyses have suggested that there is a need for other approaches to encouraging the consideration of the resulting information and adoption of risk management strategies. Hence, there is more interest in risk communication and considerations of how technical information can serve policy ends.

There has, however, been something of a countervailing development in political economy, questioning the role of governments in risk management, due to the inherent infringements on individual freedom and the public expenditure. The debates on drought policy in both the United States and Australia reflect this tendency. The argument for greater self-reliance is stronger when drought policy is focussed on agriculture, as has been the case in Australia, since this is largely about private businesses, but broader consideration of drought impacts opens up other policy demands. Chapter 4 is a review of efforts to shift the responsibility for managing the risks of agricultural droughts back to farmers in Australia. The next chapter puts a case for considering the broader effects of drought, including on nonhuman species, all types of water users, and rural residential areas, with particular discussion of the implications of climate change for a range of policy areas.

References

Adams, J. 1995. *Risk*. London: UCL Press Limited.

Aitkin, D. 1985. 'Countrymindedness'—The spread of an idea. *Australian Cultural History* 4:34–41.

Axelrod, L. J., T. McDaniels, and P. Slovic. 1999. Perceptions of ecological risk from natural hazards. *Journal of Risk Research* 2 (1): 31–53.

Beck, U. 1999. *World risk society*. Cambridge: Polity Press.

———. 2009. World risk society and manufactured uncertainties. *European Journal of Philosophy & Public Debate* 1 (2): 291–299.

Beck, U., W. Bonss, and C. Lau. 2003. The theory of reflexive modernization: Problematic, hypotheses and research programme. *Theory, Culture & Society* 20 (2): 1–33.

Botterill, L. C. 2006. Soap operas, cenotaphs and sacred cows: Countrymindedness and rural policy debate in Australia. *Public Policy* 1 (1): 23–36.

———. 2009. The role of agrarian sentiment in Australian rural policy. In *Tracking rural change: community, policy and technology in Australia, New Zealand and Europe*, ed. F. Merlan and D. Raftery, 59–78. Canberra: ANU ePress.

Buttel, F., and W. L. Flinn. 1975. Sources and consequences of agrarian values in American society. *Rural Sociology* 40 (2): 134–151.

———. 1977. Conceptions of rural life and environmental concern. *Rural Sociology* 42 (4): 544–555.

Calman, K. 2008. Beyond the 'nanny state': Stewardship and public health. *Public Health* 123:e6–e10.

Clarke, S. 2002. A definition of paternalism. *Critical Review of International Social and Political Philosophy* 5 (1): 81–91.

Dalecki, M. G., and C. M. Coughenour. 1992. Agrarianism in American society. *Rural Sociology* 57 (1): 48–64.

Ekberg, M. 2007. The parameters of the risk society: A review and exploration. *Current Sociology* 55 (3): 343–366.

Fischhoff, B. 1998. Risk perception and communication unplugged: Twenty years of process. In *Risk & modern society,* ed. R. Lofstedt and L. Frewer, 134–145. London: Earthscan Publications Ltd.

Fischhoff, B., and J. Kadvany. 2011. *Risk: A very short introduction.* Oxford: Oxford University Press.

Flinn, W. L., and D. E. Johnson. 1974. Agrarianism among Wisconsin farmers. *Rural Sociology* 39 (2): 187–204.

Giddens, A. 1999. Risk and responsibility. *The Modern Law Review* 62 (1): 1–10.

Gostin, L. O., and K. G. Gostin. 2009. A broader liberty: J. S. Mill, paternalism and the public's health. *Public Health* 123:214–221.

Jochelson, K. 2006. Nanny or steward? The role of government in public health. *Public Health* 120:1149–1155.

Lofstedt, R. 2005. *Risk management in post-trust societies.* Houndmills: Palgrave Macmillan.

Montmarquet, J. A. 1989. *The idea of agrarianism.* Moscow, ID: Idaho University Press.

Moss, D. 2002. *When all else fails: Government as the ultimate risk manager.* Cambridge, MA: Harvard University Press.

Po, M., J. Kaercher, and B. Nancarrow. 2003. *Literature review of factors influencing public perceptions of water reuse.* Perth: CSIRO Land and Water.

Renn, O. 2008. *Risk governance.* London: Earthscan Ltd.

Shrader-Frechette, K. 1998. Scientific method, anti-foundationalism and public decision making. In *Risk & modern society,* ed. R. Lofstedt and L. Frewer, 45–55. London: Earthscan Publications Ltd.

Sjoberg, L. 1998. Explaining risk perception: An empirical evaluation of cultural theory. In *Risk & modern society,* ed. R. Lofstedt and L. Frewer, 115–131. London: Earthscan Publications Ltd.

Skogstad, G. 1998. Ideas, paradigms and institutions: Agricultural exceptionalism in the European Union and the United States. *Governance* 11 (4): 463–490.

Slovic, P. 1987. Perception of risk. *Science* 236:280–285.

———. 1998. Perception of risk. In *Risk & modern society,* ed. R. Lofstedt and L. Frewer, 31–43. London: Earthscan Publications Ltd.

Wildavsky, A., and K. Dake. 1998. Theories of risk perception: Who fears what and why? In *Risk & modern society,* ed. R. Lofstedt and L. Frewer, 101–113. London: Earthscan Publications Ltd.

Willits, F. K., R. C. Bealer, and V. L. Timbers. 1990. Popular images of 'rurality': Data from a Pennsylvania survey. *Rural Sociology* 55 (4): 559–578.

3

The Science and Policy of Climate Variability and Climate Change: Intersections and Possibilities

Geoff Cockfield and Stephen Dovers

CONTENTS

Drought policies have generally been attempts to manage, adapt to, and ameliorate the impacts of climate variability with the assumption that this is variability around long-term averages in rainfall, temperatures, and crop and water yields. As noted in Chapter 1, this results in either or both reactive policies to address the effects of particular droughts and anticipatory policies to encourage resources managers to prepare for dry periods. With the emergence of concerns about climate change, discussions of drought policy will increasingly have to consider both variability and the possibility of underlying long-term changes in the key climatic variables—that is, nonstationarity. This chapter will first review the points where the sciences relating to climate change and climate variability are intersecting. Second, the policy implications of adjusting to increased variability and, in some cases, increasing drought frequency are considered. Third, there is a discussion of the ways in which popular narratives about drought and climate change interact, with a particular focus on the ways that resource users interpret the two phenomena and in some cases use one (drought) to refute the other (climate change).

There are four main contentions in this chapter:

1. Climate change modelling and seasonal variability modelling will need to be integrated so that policy makers are aware of possible changes in the frequency, length, and severity of droughts, and this is underway.

2. Related to the previous point, policy makers face a dilemma in deciding how to deal with the prospect of regions that experience those increases in drought frequency, length, and severity. That is, will the old standards of drought declaration (see Chapters 1, 4, and 9) remain, leading to increased public expenditure, or will governments try to reset expectations so that resources managers carry more of the risk in a changing climate? (See Chapter 2.)

3. Projections of climate change have implications for a number of other climate-driven perturbations impacting on society as well as drought (flood, bushfire, cyclones, storms), in terms of future imperatives for science and policy. The interaction or cumulative effects of these will increase demands on policy and public and private resources in ways that may have implications for a specific disturbance category such as drought.

4. This integration of climate change and climate variability, and policy issues other than drought, will add considerably to the complexity at the science/policy interface. There are three factors in particular that contribute to this complexity. Foremost are the uncertainties related to the extent of global warming and to what extent it will affect climate variability and resources conditions. There are also uncertainties over the efficacy of policy interventions aimed at better enabling adaptation to climatic variability and change. Then there is the breadth of the effects to be managed, especially considering the broader view of drought set out in Chapter 1. In addition to changes in regional agricultural production patterns, policy makers will need to consider the possibility of increasingly constrained water supplies, threats to production and conservation forests, and an increased threat of bushfires (wildfires).

Finally, while policy making for climate variability is inevitably a political process (see Chapters 4–9), the conflict may escalate with integration of climate change science to the policy arena. At the international and national levels, climate change as an idea is already caught up in a larger ideological contest and, at the local level, the effects of climate change may appear to pose an existential threat to the way resource-dependent communities live. So, scientists who have long focussed on climate variability and the identification of factors influencing seasonal and perhaps interdecadal outcomes are to be drawn into a more political, and perhaps more hostile, science–policy interface and will be competing with many more policy actors. To illustrate some of the issues at the interface, we use examples from Australia, particularly focussing on regions that are expected to have more low rainfall years and higher temperatures into the future.

Integrating Climate Variability and Change

Traditionally, drought research has been strongly focussed on seasonal climate forecasting (Nelson et al. 2008, 8) but from the 1990s and more extensively from the mid-2000s, there have been attempts to model the effects of climate change on climate variability (see Kirono, Hennessy, et al. 2011, 2–9, for a summary of some studies). The general approach is to develop climate change scenarios and then incorporate these into drought models, usually based on indices, to create scenarios of future drought patterns (Kirono, Hennessy, et al. 2011, 5). The climate change scenarios are developed from global climate models, using several (usually three) different gas concentration/warming patterns, and then simulating changes through the twenty-first century, especially focussing on the snapshot years of 2030 and 2070. To examine the effects of climate change on drought, a base period—for example, 1900–2007 (Kirono, Hennessy, et al. 2011, 12)—is selected and then years in that base period are classified according to indicator (temperature, rainfall, etc.) thresholds to identify 'drought' years. Then, through one or more of several possible methods (Kirono, Hennessy, et al. 2011, 14–20), the historical data are projected forward with the projected climate change assumptions added to the model.

From the modelling, five effects of climate change on drought events have been proposed:

- Increased frequency of droughts (Mpelasoka et al. 2008; Dai 2011, 56; Wang et al. 2011)
- Increased duration (Gutzler and Robbins 2011; Wang et al. 2011)
- Increased intensity (Gutzler and Robbins 2011)
- Longer recovery periods (Gutzler and Robbins 2011, 847)
- Globally, a greater proportion of land area under drought (Burke and Brown 2008)

The main contributors to global drying are Africa, southern Europe, east and south Asia, and eastern Australia (Dai 2011, 59). The modelling tends to suggest drier conditions in the subtropical regions (Seager et al. 2007, 1181) and, relevant to this study, notably in the southwest of the United States (Seager et al. 2007; Gutzler and Robbins 2011, 835) and in much of Australia, especially the southern and western regions of the mainland (Hennessy et al. 2008; Kirono, Kent, et al. 2011; overview in Kirono, Hennessy, et al.

2011). The uncertainties from the integrated modelling will be discussed after consideration of the implications for policy making if there is to be an increase in the frequency, intensity, and duration of droughts, at least in some regions.

The Policy Implications

If droughts, as they are now defined, become more frequent, more intense, and longer, then this will put pressure on governments to change policy approaches in several areas. Most obviously, there will be a challenge in considering the management of agricultural impacts if past drought events are used as a benchmark. Second, there will be increasing pressure on water supply so that the distribution of water, especially the allocation between agricultural and other uses, will be made more difficult. Then there are the implications for nonagricultural land uses, such as for production and non-production forests, including those established either fully or partially for carbon sequestration. Increasingly frequent and severe droughts, along with higher temperatures, could influence where timber production will occur in the future and the location, composition, and environmental values of conservation forests and remnants.

Agricultural Policy

The primary potential socioeconomic impacts of climate change on some agricultural regions, excluding for the moment adaptations and productivity gains, will be an increase in the variability of farm incomes and an overall decrease in average incomes. To illustrate this, some researchers have already linked rainfall patterns in the southwest of Western Australia (WA) to global warming (Hennessy et al. 2008, 3). The inland areas of this region are dominated by wheat production based on winter rainfall, but there has been an increase in the number of low rainfall years (Hennessy et al. 2008, 8; 10). Table 3.1 shows the rainfall results from the wheat belt town of Wagin, divided into 30-year periods. The number of years where rainfall is very much below the cropping season average (375 mm) by 100 mm or more has increased in each period, while the number of 'wet' years (100 mm or more above the average) has decreased.

If this were the trend into the future, aside from productivity gains which have been characteristic of WA wheat production, yields and income would likely become more variable and, on average, perhaps even decline. This could eventually mean that farm values should decrease, reflecting the lower returns on investment. These outcomes would leave current and future Australian governments with two broad choices, if they want to control

TABLE 3.1

Proportion of Years by Difference from the Average Winter Crop Season
Rainfall at Wagin[a] (WA)

	Period (Years)			
Difference from the Average[b]	**1891–1820**	**1921–1950**	**1951–1980**	**1981–2010**
100 mm or more below average	3.3	10.0	16.7	23.3
50 to 100 mm below average	13.3	13.3	10.0	16.7
1 to 50 mm below average	30.0	10.0	30.0	26.7
0 to 50 mm above average	23.3	23.3	23.3	13.3
50 to 100 mm above average	13.3	23.3	6.7	20.0
100 mm or more above average	16.7	20.0	13.3	0.0

Source: Bureau of Meteorology 2011.

[a] There were two missing monthly results, which were filled by records from the nearest station.

[b] The winter crop season is counted as March to October, with planting from May to June.

expenditure and encourage farmers to adapt. They could implicitly normalise the increase in the number of lower rainfall years by keeping in place, and perhaps more heavily weighting, the current criteria of exceptional years being 1 in 25 (see Chapter 6 for elaboration). Over time, the increasing number of dry years would effectively increase the threshold before intervention was triggered. This has some obvious drawbacks. It would take time for the moving averages to shift the thresholds significantly. Furthermore, an absence of explicit discussion of an effective shift in policy would seem underhanded and could work against the provision of adaptation advice. Finally, it is doubtful that governments would have the political will to maintain the line. Drought declarations in Australia have, since 2000, routinely exceeded the criteria set for the National Rural Advisory Council and the Minister (Nelson et al. 2008, 594).

The second approach would be explicitly to normalise the increased frequency of low rainfall years by openly raising the thresholds before government support was triggered and to provide some offsetting programmes to facilitate farm and regional level adaptation. These programmes could include additional funding for agronomic research to facilitate changes in production techniques, including biotechnology and structural adjustment schemes. Australian farmers have already shown a capacity to deal with variability in that those regions where there is the highest rainfall variability are not necessarily those that have the highest variability in income (Nelson, Kokic, Crimp, Martin, et al. 2010, 21). Australia also has a long history of adjustment schemes for agriculture (for a review, see Cockfield and Botterill 2006). These schemes do not necessarily involve a high demand from farmers, who show considerable inclination to make their own adjustments, but they are politically important (Cockfield and Botterill 2006). Extending this approach, vulnerable (to climate change) regions could be identified and this could be the basis for time-limited regional adjustment programmes, as have been used for parts of the pastoral zone and for particular industries, such as dairying. Scientists will be drawn into this area of policy in developing the vulnerability maps and this could be a fraught exercise.

Water Policy

Possibly even more controversial will be water management, given the increasing demands from a growing population and for those seeking environmental flows (Mpelasoka et al. 2008, 1284). This conjunction of supply and demand problems is evident in the Murray–Darling Basin (MDB), where relatively low inflows and high variability would be compounded by climate change (Adamson et al. 2009). The MDB contains all the major Australian irrigation areas but, as with the WA wheat belt, there was an illustration of the future when the federal government virtually eliminated water allocations in the southern areas in 2007, during the millennium drought (2002–2009)—effectively suspending rice production in Australia. The other impact on supply is the programme to increase environmental flows, with the federal government 'buying' back water based on regional targets that were initially proposed to retract 20–49 percent of current diversions* from the Basin (Murray–Darling Basin Authority 2010, xix). While the scientific basis of proposed water allocation changes in the Basin is contested, the social science evidence base regarding impacts of changes is just as important in a policy and political sense (see Chapter 8), an issue made apparent in community backlash against the earlier iterations of the Murray–Darling Basin Plan (Crase et al. 2011). The science–policy interface therefore does and must include other forms of 'science' in addition to the hydrological and meteorological disciplines.

Australia has also adopted a partial market-based approach to water management, and water can be traded within the MDB, including for purposes other than irrigation (National Water Commission 2010). The effects of this are likely to be that some of the production-smoothing effects of irrigation schemes, which were one of the main reasons for their development, will be negated. In drought years, spot (temporary sales) water prices will increase so that water will be traded away from crop production or costs of production will increase for those buying water. This will affect regional incomes, furthering the impact of policy change beyond irrigators. It should be noted that regional development was another of the main reasons for establishing irrigation schemes in the first place. Over time, water allocations will increasingly move to higher value uses, including for mining, boutique crops such as wine, and domestic consumption, changing agricultural landscapes.

---

* Subsequent political unrest is likely to mean such targets will be modified, especially in light of higher rainfall in the MDB in 2010 and 2011.

The policy dilemmas for governments include deciding the scale of environmental flows, whether or not to hold the line on environmental flows during dry years, and whether or not to provide regional adjustment programmes where irrigation decreases significantly within a region. More boldly, federal and state governments could decide to return to 'nation building' and support or coordinate new or expanded irrigation developments in the northern regions of Australia, where rainfall may not be so significantly reduced under climate change. There was considerable political interest in this approach at the height of the millennium drought (Shanahan 2007; ABC News 2009), as well as some more measured scientific and socioeconomic analyses of the possibilities (Camkin et al. 2007; Petheram et al. 2008; Alexander and Ward 2009; Northern Australia Land and Water Taskforce 2009). The potential of northern agricultural development is, however, highly contested (and has been for decades) and may not be a replacement for significant lost production in southern areas.

An increase in the frequency and severity of drought will also require responses in relation to the management of urban water storages. Australia is highly urbanised with more than 60 percent of people living in the five major (state capital) cities. Drought years have historically been managed by having sufficient storage to cover most years with escalating water restrictions as the rainfall deficit increases above a threshold (England 2009). Dam construction slowed considerably from the 1980s as governments tried to limit expenditure and avoid clashes over the environmental consequences of damming waterways. However, as in rural areas, convenient and cheap dam sites are no longer common. There has also been some reluctance to expose domestic and industrial consumers to the same market forces as irrigators (Brennan 2008; Crase et al. 2009). Climate change projections suggest a decrease in rainfall and runoff, and hence a long-term decrease in rain-fed supply to current storages (Brennan 2008, 8–10). Alternative supply strategies include grey water reuse and household tanks (England 2009, 603), with rainwater tanks considered to be more acceptable than grey water reuse, the former being something of an Australian tradition (Po et al. 2003, 21). Such decentralised systems rely, however, on a myriad of individuals to install and maintain equipment and there are some health concerns with household tanks (Beebe et al. 2009; Ahmed et al. 2010).

Some authors suggest the need for a significant recasting of urban water supply and use to adapt more strongly to variability in climate, with a departure from a long-standing reliance on large-scale engineered solutions (see various authors in Troy 2008), and others suggest that recent urban water policy trends are, in effect, a 'maladaptation' to climate change (Barnett and O'Neill 2010). One heavily criticised adaptation being adopted in Australian cities is desalination, which is virtually drought proof (El Salibya et al. 2009, 2) but would increase power costs and greenhouse gas emissions (El Salibya et al. 2009, 5) as well has having some other in-sea environmental concerns (Roberts et al. 2010). Recycling waste water is another option but water use involving high levels of human contact is persistently unpopular with the public (Po et al. 2003; Marks et al. 2008; Dolnicar and Hurlimann 2010).

The concerns about both urban and irrigation water during drought highlight the need for greater coordination between urban and rural water policy and management, and the evidence base for policy in each domain; as to date, these have largely been separate in Australia. The 2004 National Water Initiative (Council of Australian Governments 2004), the key overarching water policy, aims to drive integration of urban and rural systems. Increasing droughts would be expected to increase demands for water transfers between regions and there will be tension over specific proposals and contestation over scientific and other information used to support or oppose developments. As an example, proposals to deliver water via pipeline from rural Victoria to the city of Melbourne met considerable opposition from rural water users, who considered this akin to removing a regional resource.

Once again, taking a broad view of the consequences of drought and the related policy communities, water availability and use will be a critical interface of science and policy, in a highly political environment. For both agricultural and domestic water supply, there will be the usual hydrometereological modelling, work on water delivery efficiency, and ecological assessments of the impacts of dams and environmental flows. In regard to irrigated agriculture, the roles for the various types of scientists will include developing hydrometeorological-socioeconomic models to aid farm decision making (Adamson et al. 2009), advising on improvements in water use efficiency, and undertaking agronomic research for irrigated crop production. In regard to domestic water supply there could be, depending on the political will and inclination of governments, research and extension on water recycling, desalination and power generation for desalination, and the efficacy and safety of household rainwater supplies. The evidence base for choosing between such options is at best contested, and more often weak.

Nonagricultural Land Use Policies

Droughts also have an effect on other land uses, notably on production and conservation forests and rural residential areas. Apart from driving up the costs of domestic water in rural residential (or rural lifestyle) areas, droughts can contribute to the bushfire threat. It is expected that, in Australia, climate change will lead to an increase in the frequency and intensity of bushfires (Flannigan et al. 1998; Flannigan et al. 2000), and if droughts are more frequent, longer, and more severe, as proposed previously, then the threats will escalate accordingly. This will lead to an increase in fire-fighting costs and/or fire-prevention work and/or the regulation of rural residential settlement patterns.

The increased fire risk also has ramifications for conservation land managers. Increased frequency and intensity of fires could change species composition and ecosystems' processes (Hughes 2003) and accelerate the changes to ecosystems resulting from climate change (see Cayan et al. 2008). As argued by Aber et al. (2001), Dale et al. (2001), and McNulty and Abe (2001), future climate change will significantly affect the distribution, condition, species composition, and productivity of forests. Droughts will also affect the growth rates of both timber and 'sequestration' forests. Sequestration forests are those that will be planted as offsets for greenhouse gas emissions (for scenarios, see Lawson et al. 2008; Burns et al. 2009).

Going beyond the traditional focus of drought policy (farming) in a potentially more drought-affected world with changed climate involving increased fire risk will mean that the complexities of science, policy, and politics become significantly greater. In some regions, there will be intersection between multiple land uses: agriculture (of multiple kinds), rural, and other demands for water; conservation in and outside the reserve system; production forests; and carbon forests. To the traditional hydrological and meteorological science informing drought policy will be added more sophisticated hydrological work characterising forest water use, landscape ecology informing connectivity conservation (Worboys et al. 2010), socioeconomic evidence regarding impacts of land use change, carbon accounting, and fire science. None of these is immune from contestation in a scientific sense—let alone political sense—and integration of them in strategic land use decision making will be highly complicated. For example, debates over fuel reduction (prescribed) burning continue, especially since the 5 percent of landscape target was promulgated by the Royal Commission into the devastating 2009 Victorian bushfires, while strong evidence of the efficacy or not of such treatments is incomplete and only recent (Gibbons et al. 2012).

Climate science, broadly defined, concerns much more than drought. Sharper expressions of climate change and increased variability will be encountered across many areas of emergency management, including heat waves, floods, cyclones, and storms. The challenges mounted by climate change to emergency policy and management are potentially serious (Handmer and Dovers 2007), particularly if cumulative or more sequential events occur. For drought policy, the available policy and financial resources and scientific attention may be diluted in a future where various deleterious climatic events occur more often.

Uncertainty and Complexity at the Science–Policy Interface

For scientific information to be usable in business and policy decision making, it must be credible (seen as technically accurate), salient (seen to be relevant), and legitimate (not just supporting a narrow agenda) (Meinke et al. 2006, 102; Stone and Meinke 2006). Other chapters will review the performance of the traditional research and extension of climate variability sciences. The purpose here is to suggest that the integration of research into climate variability and climate change will be very difficult on all three criteria. Climate change has become part of a political debate in which there is particular interest in undermining the *legitimacy* of the underlying science. If end users of the information, such as farmers, do not believe that there is an effect or that it has any connections to drought, then the hybrid modelling has no *relevance*. Finally, the *credibility* of the climate change research is much more easily questioned than that for climate variability because of the high degree of uncertainty. These challenges for science become larger when different forms of information must be integrated in judging multiple land uses and comparing multiple policy options with different impacts on sectors of society.

The uncertainties include the degree of global warming that will occur, the effects on the climate, and how particular regions will be affected (Hennessy et al. 2008; Kirono, Hennessy, et al. 2011). The modelled effects depend on which warming scenario and which climate change models are used (Wang et al. 2011, 9). There is also uncertainty as to whether or not conventional indicators of drought will be useful measures under climate change (Dai 2011, 60). Then there are the issues of legitimacy and relevance. While the majority of Australians in one survey believe climate change is happening (more than 80 percent), only half believe it is anthropogenic (Leviston and Walker 2011, 5). In another survey, 74 percent believe the climate is changing but less than one-third attribute this mainly or entirely to anthropogenic causes (Reser et al. 2010, 16–17). Of particular relevance to the discussion of agricultural impacts, there are differences between urban and farm folk in regard to the belief in anthropogenic causation, with 58 percent of urban people thinking so compared to 27 percent of rural respondents (Donnelly et al. 2011, 14). More than half of a sample of 1,000 farmers (including primary and secondary decision makers) thought that drought would not be made worse by climate change, compared to only one-quarter of the urban sample (also 1,000 people) (Donnelly et al. 2011, 14).

There are understandable psychological reasons for landholders to be reluctant to accept climate change, as it could be seen as an existential threat (Donnelly et al. 2011, 19). Farmers also place a high level of importance on place experience and can point to past 'cycles' as evidence that long dry periods, such as the Australian millennium drought, have 'happened before' (Donnelly et al. 2011, 18). They also tend to highlight the supposed divisions amongst scientists in regard to climate change (Donnelly et al. 2011, 18), just as those more generally who do not believe in climate change tend to have lower mean ratings of trust of scientists than those who do (Leviston and Walker 2011, 8). This distrust may be fuelled by media: selection of stories, the search for conflict (promoting scepticism), political influence, and the interpretation of uncertainty (Speck 2010). There is also evidence of contrarians gearing up from the early 1990s to influence media coverage (Boykoff and Roberts 2007, 24).

Climate change projections and their implications are not the only source of uncertainty relevant here, as uncertainties attach to the science relevant to many other future land use and resource allocation scenarios raised here including carbon uptake, vegetation water use, and socioeconomic impacts. Furthermore, when drought policy is considered within the more complex space of climate adaptation, considerably greater uncertainty attends the efficacy of competing or alternative policy options (Dovers and Hezri 2010). While there is a reasonably robust language of uncertainty used in climate science regarding the availability of *evidence* and degree of *confidence* (IPCC 2007), there is no such nomenclature available for alternative policy options (for example, market or insurance solutions versus support payments for drought, or educative versus regulatory measures for bushfire preparedness) other than political claim. Further uncertainty attaches to future social preferences and political trends, which are of crucial importance to future policy styles and choices above and beyond any scientific evidence of the need for change.

Given these uncertainties and that climate change science and policy will remain embroiled in broader political conflict about environmentalism, growth, business obligations, and the role of government, we see the necessity for adaptive management in translating the science into policy. There is a continuum of science/policy engagement from 'centralised expert management' to 'adaptive governance' (Nelson et al. 2008, 589). In environmental matters, centralised expert management has been pushed by the idea that there need to be external drivers to break the cycle of resource exploitation by resource holders (Nelson et al. 2008, 589). Critics see this as resulting in reductionist science with the setting of thresholds and so on. This leads to an over-reliance on quantitative measurement and analysis and it is backward looking (Nelson et al. 2008; Nelson, Kokic, Crimp, Meinke, et al. 2010). This reductionism tends to favour national approaches and thresholds and it requires policy makers to be clear on goals, whereas this is rarely the case. As a result,

> Climate policy in Australian agriculture, including drought policy, has tended to focus on the management of climate risk within existing farming systems and rural livelihood strategies. This approach...is at risk of being stranded by the uncertainty surrounding climate change, because the maintenance of existing farming systems and livelihood strategies may no longer be possible. Consequently, there is an urgent need to replace traditional climate risk management with adaptive approaches better suited to managing uncertainty. (Nelson et al. 2008, 593)

This adaptive governance would also require a degree of devolution to include regional and local institutions and conditions (Nelson et al. 2008, 597). This would be in parallel with more 'rule of thumb' approaches such as some form of carrying capacity (2008, 597), which takes account of local conditions. The extension of this would be localised data collection and management. The science would move to interpretive measures (2008, 598).

However, there is little agreement or specificity regarding what 'adaptive' policy measures or (even more so) governance arrangements might look like. It has been noted that the literature on resilience, adaptive governance, and climate adaptation are not well connected to highly relevant areas such as public policy and administration or institutional theory (Dovers and Hezri 2010). At least four considerations arise:

1. The political and legal implications of adaptive policy regimes—that is, contingency, uncertainty, and flexibility (change)—are far from clear.

2. Establishing flexible or adaptive policy settings that can persist but be flexible over long time frames (in comparison to political and policy cycles) is something that few jurisdictions evidence as a mature ability.

3. Precisely what policy instruments suit 'adaptiveness' is unclear, across the wide menu of regulatory, self-regulatory, community-based, informational, market-based, etc., and what governance regimes do. An exploration of relevant Australian policy sectors against the theme of 'adaptive policies and institutions' used five interacting principles—persistence, purposefulness, information richness and sensitivity, inclusiveness, and flexibility—and indicated the context-specific and complex nature of achieving an adaptive regime (Dovers and Wild River 2003). In the context of drought and other climate-sensitive issues, should we favour the *persistence* and *purposefulness* of clear regulation, the *inclusiveness* of local decision making, the *flexibility* of markets, the *information sensitivity* of a science-based threshold approach, or some combination of these?

4. The call for devolution raises long argued issues around the location of decision making, statutory competence, and information generation in a federal political system, especially in the Australian rural domain where a fourth, regional level of governance, the catchment management authorities, has been created and embraced but given highly circumscribed authority (Lane et al. 2009). Should an adaptive approach as promoted by Nelson et al. (2008) be read as implying a continuation of the contested, difficult, and protracted move to local self-reliance at property and local community scale described elsewhere in this book, but in an even more difficult policy environment?

Conclusions

It is clear that the difficult intersection between science and policy in the drought policy domain will not be resolved anytime soon. It is highly likely that it will in fact be more contested and complicated, and the four contentions raised at the start of this chapter seem to hold. First, there is a need for integration of climate change and seasonal variability modelling—a task for scientists and a challenge for policy makers then to utilise. Second, the policy dilemma following acceptance of the likelihood of climate change thus characterised is real: Should public policy and the public purse seek to meet increased expectations, or must these expectations be lowered, suggesting a different policy approach? Third, drought policy cannot be considered alone when considering science and policy interactions under changing climatic conditions, but rather is closely related to natural resource management and other policy areas. Finally, and following from the preceding, there is little reason to expect that the science–policy interface will become anything other than more demanding, complex, and contested.

References

ABC News. 2009. Rudd touts Ord River expansion as 'new food basket'. ABC, December 16 2008 [Accessed January 10, 2009]. Available from http://www.abc.net.au/news/stories/2008/12/16/2447857.htm

Aber, J., R. P. Neilson, S. McNulty, J. M. Lenihan, D. Bachelet, and R. J. Drapek. 2001. Forest processes and global environmental change: Predicting the effects of individual and multiple stressors. *BioScience* 51:735–751.

Adamson, D., T. Mallawaarachchi, and J. Quiggin. 2009. Declining inflows and more frequent droughts in the Murray–Darling Basin: Climate change, impacts and adaptation. *The Australian Journal of Agricultural and Resource Economics* 53:345–366.

Ahmed, W., A. Vieritz, A. Goonetilleke, and T. Gardner. 2010. Health risk from potable and non-potable uses of roof-harvested rainwater in Australia using quantitative microbial risk assessment. *Applied Environmental Microbiology* 76:7382–7391.

Alexander, K., and J. Ward. 2009. The current status of water governance in northern Australia: Water management in the Northern Territory, Queensland and Western Australia. In *Northern Australia land and water science review full report*, ed. P. Stone, 20.1–20.49. Canberra: CSIRO.

Barnett, J., and S. O'Neill. 2010. Maladaptation. *Global Environmental Change* 20:211–213.

Beebe, N., R. Cooper, P. Mottram, and A. Sweeney. 2009. Australia's dengue risk driven by human adaptation to climate change. *PLoS Neglected Tropical Diseases* 3 (5): 1–9.

Boykoff, M. T., and J. T. Roberts. 2007. Media coverage of climate change: Current trends, strengths, weaknesses. Human Development Report Occasional Paper: United Nations Development Programme.

Brennan, D. 2008. Will we all be Rooned without a Desal plant? Hanrahan's lament and the problem of urban water planning under climate change. *Agenda* 15 (3): 5–20.

Bureau of Meteorology. 2011. Monthly rainfall Wagin climate data online: Bureau of Meteorology. http://www.bom.gov.au/jsp/ncc/cdio/weatherData/av?p_nccObsCode = 139&p_display_type = dataFile&p_stn_num = 010647

Burke, E. J., and S. J. Brown. 2008. Evaluating uncertainties in the projection of future drought. *Journal of Hydrometeorology* 9:292–299.

Burns, K., J. Vedi, E. Heyhoe, and H. Ahammad. 2009. *Opportunities for forestry under the Carbon Pollution Reduction Scheme (CPRS): An examination of some key factors.* Canberra: Australian Bureau of Agricultural and Resource Economics.

Camkin, J., B. Kellett, and K. Bristow. 2007. *Northern Australia irrigation futures: Origin, evolution and future directions for the development of a sustainability framework.* Townsville: CRC for Irrigation Futures and CSIRO.

Cayan, D. R., A. L. Luers, G. Franco, M. Hanemann, B. Croes, and E. Vine. 2008. Overview of the California Climate Change Scenarios Project. *Climatic Change* 87 (1): S1–S6.

Cockfield, G., and L. C. Botterill. 2006. Rural adjustment schemes: Juggling politics, welfare and markets. *Australian Journal of Public Administration* 65 (2): 70–82.

Council of Australian Governments. 2004. *Intergovernmental agreement on a national water initiative.* Canberra: CoAG (www.coag.gov.au/meetings/250604/).

Crase, L. R., S. M. O'Keefe, and B. E. Dollery. 2009. The fluctuating political appeal of water engineering in Australia. *Water Alternatives* 2 (3):441–447.

———. 2011. Some observations about the reactionary rhetoric circumscribing the guide to the Murray–Darling Basin plan. *Economic Papers* 30 (2): 195–207.

Dai, A. 2011. Drought under global warming: A review. *Climate Change* 2:45–65.

Dale, V. H., L. A. Joyce, S. McNulty, R. P. Neilson, M. P. Ayres, M. D. Flannigan, P. J. Hanson, et al. 2001. Climate change and forest disturbances. *BioScience* 51:723–734.

Dolnicar, S., and A. Hurlimann. 2010. Water alternatives—Who and what influences public acceptance? *Journal of Public Affairs* DOI: 10.1002/pa.378.

Donnelly, D., R. Mercer, J. Dickson, and E. Wu. 2011. *Australia's farming future—Final market research report: Understanding behaviours, attitudes and preferences relating to climate change.* Sydney: Australian Government Department of Agriculture, Fisheries and Forestry.

Dovers, S., and S. Wild River, eds. 2003. *Managing Australia's environment.* Sydney Federation Press.

Dovers, S. R., and A. A. Hezri. 2010. Institutions and policy processes: the means to the ends of adaptation. *Wiley Interdisciplinary Reviews: Climate Change* 1 (2): 212–231.

El Salibya, I., Y. Okoura, H. Shona, J. Kandasamya, and S. In. 2009. Desalination plants in Australia, review and facts. *Desalination* 247:1–14.

England, P. 2009. Managing urban water in Australia: The planned and the unplanned. *Management of Environmental Quality* 20 (5): 592–608.

Flannigan, M. D., Y. Bergeron, O. Engelmark, and B. M. Wotton. 1998. Future wildfire in circumboreal forests in relation to global warming. *Journal of Vegetation Science* 9:469–476.

Flannigan, M. D., B. J. Stocks, and B. M. Wotton. 2000. Climate change and forest fires. *The Science of the Total Environment* 262 (3): 21–229.

Gibbons, P., L. van Bommel, A. M. Gill, C. J. Cary, D. A. Driscoll, R. A. Bradstock, E. Knight, M. A. Moritz, S. L. Stephens, and D. B. Lindenmayer. 2012. Land management practice associated with house loss in wildfires. *PLoS One* 7 (1):e29212 DOI: 10.1371.journal.pone.0029212.

Gutzler, D., and T. Robbins. 2011. Climate variability and projected change in the western United States: Regional downscaling and drought statistics. *Climate Dynamics* 37:835–849.

Handmer, J., and S. Dovers, eds. 2007. *The handbook of disaster and emergency policies and institutions.* London: Earthscan.

Hennessy, K., R. Fawcett, D. Kirono, F. Mpelasoka, D. Jones, J. Bathols, P. Whetton, M. Stafford Smith, M. Howden, C. Mitchell, and N. Plummer. 2008. *An assessment of the impact of climate change on the nature and frequency of exceptional climatic events.* Canberra: Bureau of Meteorology and CSIRO.

Hughes, L. 2003. Climate change and Australia: Trends, projections and impacts. *Austral Ecology* 28:423–443.

IPCC. 2007. Summary for policymakers. In *Climate change 2007: The physical science basis. Contribution of Working Group I to the Fourth Assessment Report of the Intergovernmental Panel on Climate Change,* ed. S. Solomon, D. Qin, M. Manning, Z. Chen, M. Marquis, K. B. Averyt, M. Tignor, and H. L. Miller. Cambridge: Cambridge University Press.

Kirono, D. G. C., K. Hennessy, F. Mpelasoka, and D. Kent. 2011. Approaches for generating climate change scenarios for use in drought projections—A review. CAWCR Technical Report. The Centre for Australian Weather and Climate Research.

Kirono, D. G. C., D. M. Kent, K. J. Hennessy, and F. Mpelasoka. 2011. Characteristics of Australian droughts under enhanced greenhouse conditions: Results from 14 global climate models. *The Journal of Arid Environments* 75:566–575.

Lane, M., C. Robinson, and B. Taylor, eds. 2009. *Contested country: Local and regional natural resources management in AustrLawson, K., K. Burns, K. Low, E. Heyhoe, and H. Ahammad. 2008. Analysing the economic potential of forestry for carbon sequestration under alternative carbon price paths.* Canberra: Australian Bureau of Agricultural and Resource Economics.

Leviston, Z., and I. A. Walker. 2011. Baseline survey of Australian attitudes to climate change: Preliminary report: CSIRO.

Marks, J., B. Martin, and M. Zadoroznyj. 2008. How Australians order acceptance of recycled water: National baseline data. *Journal of Sociology* 44 (1): 83–99.

McNulty, S. G., and J. D. Abe. 2001. United States national climate change assessment on forest ecosystems: An introduction. *BioScience* 51:720–722.

Meinke, H., R. Nelson, P. Kokic, R. Stone, R. Selvaraju, and W. Baethgen. 2006. Actionable climate knowledge: from analysis to synthesis. *Climate Research* 33:101–110.

Mpelasoka, F., K. Hennessy, R. Jones, and B. Bates. 2008. Comparison of suitable drought indices for climate change impacts assessment over Australia towards resource management. *International Journal of Climatology* 28:1283–1292.

Murray–Darling Basin Authority. 2010. *Guide to the proposed Basin Plan.* Canberra: Murray–Darling Basin Authority.

National Water Commission. 2010. *The impacts of water trading in the southern Murray–Darling Basin: An economic, social and environmental assessment.* Canberra: National Water Commission.

Nelson, R., M. Howden, and M. Stafford Smith. 2008. Using adaptive governance to rethink the way science supports Australian drought policy. *Environmental Science and Policy* 11:588–601.

Nelson, R., P. Kokic, S. Crimp, P. Martin, H. Meinke, S. M. Howden, P. de Voil, and U. Nidumolu. 2010. The vulnerability of Australian rural communities to climate variability and change: Part II—Integrating impacts with adaptive capacity. *Environmental Science and Policy* 13:18–27.

Nelson, R., P. Kokic, S. Crimp, H. Meinke, and S. M. Howden. 2010. The vulnerability of Australian rural communities to climate variability and change: Part I—Conceptualising and measuring vulnerability. *Environmental Science and Policy* 13:8–17.

Northern Australia Land and Water Taskforce. 2009. *Northern Australia Land and Water Science Review 2009*. Canberra: Northern Australia Land and Water Taskforce. http://www.nalwt.gov.au/science_review.aspx

Petheram, C., S. Tickell, F. O'Gara, K. Bristow, A. Smith, and P. Jolly. 2008. *Analysis of the Lower Burdekin, Ord and Katherine-Douglas-Daly irrigation areas: Implications to future design and management of tropical irrigation*. Townsville: CRC for Irrigation Futures and CSIRO.

Po, M., J. Kaercher, and B. Nancarrow. 2003. *Literature review of factors influencing public perceptions of water reuse*. Perth: CSIRO Land and Water.

Reser, J. P., N. Pidgeon, A. Spence, G. Bradley, A. I. Glendon, and M. Ellul. 2010. *Public risk perceptions, understandings, and responses to climate change in Australia and Great Britain: Interim report*. Griffith University, Climate Change Response Program, Queensland, Australia, and Understanding Risk Centre, Cardiff University, Wales.

Roberts, D., E. Johnston, and N. Knott. 2010. Impacts of desalination plant discharges on the marine environment: A critical review of published studies. *Water Research* 44:5117–5128.

Seager, R., M. Ting, I. Held, Y. Kushnir, J. Lu, G. Vecchi, H-P. Huang, et al. 2007. Model projections of an imminent transition to a more arid climate in south-western North America. *Science* 316:1181–1184.

Shanahan, D. 2007. Taskforce looks to clear path for food bowl's shift north. *The Australian*, May 2.

Speck, D. L. 2010. A hot topic? Climate change mitigation policies, politics, and the media in Australia. *Human Ecology Review* 17 (2): 125–134.

Stone, R., and H. Meinke. 2006. Weather, climate, and farmers: an overview. *Meteorological Applications* Supplement:7–20.

Troy, P., ed. 2008. *Troubled waters: Confronting the water crisis in Australia's cities*. Canberra: ANU ePress.

Wang, D., M. Hejazi, X. Cai, and A. J Valocchi. 2011. Climate change impact on meteorological, agricultural, and hydrological drought in central Illinois. *Water Resources Research* 47 (W09527).

Worboys, G. L., W. L. K. Francis, and M. Lockwood, eds. 2010. *Connectivity conservation management: A global guide*. London: Earthscan.

4

Drought, Climate Change, Farming, and Science: The Interaction of Four Privileged Topics

Peter Hayman and Lauren Rickards

CONTENTS

Each culture ranks particular risks as normal and others as special (Douglas and Wildavsky 1982). Given the low and erratic rainfall across much of Australia, drought could reasonably be seen as merely an unremarkable 'background' feature of the country, but it has been treated as a special risk. Since Australia was colonised by the rain-accustomed British in the late eighteenth century, the phenomenon of drought has provoked discussion and contested definitions (West and Smith 1996). One of the reasons for the special status of drought is the notion that farming is inherently a worthy

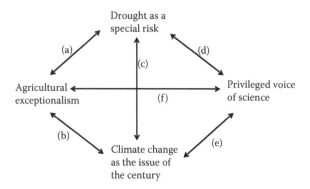

FIGURE 4.1
Drought, climate change, farming, and science.

activity, a perspective referred to in Australia as 'countrymindedness' (Aitkin 1985; Botterill 2009). Countrymindedness is one form of agrarianism and, like other such narratives in the Western world, supports arguments that the agricultural sector should rightly be treated as an exception in regard to industry and social policy (see Chapter 3). Heathcote (1973, 36) illustrates this in the following observation:

> In any catastrophe, public sympathy goes out to victims, but when those victims are the sons of the soil, on the margins of the good earth, struggling to give us our daily bread, the emotional response is tremendous and objectivity is often left behind.

Governments have often responded to the 'plight' of farmers, though some have tried to redefine drought as merely a 'background' feature.

The proponents of agricultural exceptionalism are, however, not the only voices in defining and responding to drought. Figure 4.1 shows the interaction of drought and climate change on the vertical axis and farming and science on the horizontal axis. Not only does each of these four themes hold a privileged position in the media and policy, the interaction between these topics further reinforces the privileged status of each. Even climate change, which as discussed later is also introducing new considerations and countervailing forces, has prompted responses that reflect pre-existing norms, including a keen interest in drought, concern about agriculture (both empathetic and critical), and the privileged perspective of science on the world.

The preceding quote from Heathcote refers explicitly (and slightly condescendingly) to agricultural exceptionalism. Crucially, the notion that 'objectivity is often left behind' reflects another form of exceptionalism: that afforded to science. As argued in greater detail later in the chapter, science has a privileged voice in society as a whole, including in Australian farming. The final privileged topic to highlight is that of climate change, named the century's defining issue (*Economist* 2009) or the predominant moral issue of the century comparable to Nazism faced by Churchill in the twentieth century and slavery faced by Lincoln in the nineteenth century (Hansen 2010).

In Australia, a strong relationship has been drawn between farmers as victims of a changing climate and the likelihood of stronger and more intense droughts (lines a, b, and c in Figure 4.1). Garnaut (2008) pointed out that of all developed countries, Australia has the most to lose from unmitigated climate change and that, of all sectors within Australia, agriculture is especially vulnerable. Droughts not only feature as a highly likely manifestation of climate change (Hennessy et al. 2008), but also as a reason for action. The reinforcing relationship works both ways: experiences of drought have raised the profile of climate change, and climate change has further elevated drought as a special risk because it is seen as a window into Australia's hotter and drier future (as discussed further later).

The complex relationship between farming and science (line e in Figure 4.1) is also reinforced by drought and climate change (lines c, d, and f). Farmers on the frontier of climate change present a human face to the impacts. Assisting farmers to deal with drought and a changing climate through climate science, better agronomy, or improved varieties and animal breeds not only is seen as personally satisfying to scientists, but also legitimises the role of science and may be seen as justification for funding. A specific example is the way that drought and climate change are raised in the context of the need for research on genetically modified crops; a challenge to genetic modification is then often dismissed as unscientific (Hepburn 2011).

Another way of considering Figure 4.1 is to consider alternative topics that are important but not given the same privileged status. An interesting example is the comparison between the treatment of drought and unemployment (Watts 2003). Both are associated with psychological harm, ill health, poverty (including impact on family members), and social exclusion. Furthermore, both are caused by complicated mixes of structural or systemic failure and personal responsibility. Watts (2003) argues that farming and drought are given special treatment in policy and the media, where the emphasis is on the role of external forces (namely, drought) rather than farmers' management. This is consistent with treatment of farmers as the 'deserving poor' as a manifestation of agricultural exceptionalism (Botterill 2007). Other components of Figure 4.1 can also be compared with alternatives. The natural sciences have conventionally been privileged over the social sciences in research and development for agriculture and in responding to drought and climate change more generally. Likewise, climate change receives coverage beyond other environmental stressors such as salinity, pest outbreaks, or even economic stressors such as the exchange rate. In this chapter we are not judging whether drought, climate change, farming, and science should or should not be afforded a privileged status. Rather, in the spirit of reflexivity demanded by climate change (Ison 2010), in order to help untangle current debates we are simply highlighting the way they have been and continue to be privileged.

Central to the argument of this chapter is the notion that the current interaction of farming, science, and drought risk has many historical layers. The historical horizon for many farming families and rural communities goes back generations, and this is often measured in terms of major droughts and floods and runs of good seasons and poor seasons. Contemporary discussions of drought and climate are dominated by a topic from which drought is now inseparable: anthropogenic climate change. A complex, contested, and quintessentially 'wicked' issue (Head 2008; Turnpenny and Lorenzoni 2009; Brown and Deane 2010), climate change has greatly heightened the role of science in agriculture. Reinforcing the long-standing focus on science in the agricultural sector, including on climate science and its seasonal forecasting applications, climate change inserts a quintessentially scientific issue into farming—one that is only accessible via complicated climate science. This chapter takes a historical view by identifying three eras in Australian agriculture that illustrate changing interpretations of drought risk and the enduring role of science in determining this within the sector. First, though, we consider the special but ambiguous role of drought in Australia and provide a brief overview of attempts to normalise drought.

The Special but Ambiguous Status of Drought in Australia: Getting Over the Colour Green

As a landform, livelihood, vocation, and industry, agriculture has always had a key role in the national identity of both Britons and Australians. Australia's colonial and postcolonial relationship with drought has been strongly shaped by its relationship with agriculture, and vice versa. Although it has been the focus of intense discussions over time and many farmers are adept managers of drought risk, it remains a central and perhaps growing issue within agriculture. Keogh and Granger (2011, 10) note that all involved in agriculture, including governments at state and federal levels, 'continue to grapple with the best way to manage the risks associated with drought'.

Australians have long been both frustrated and fascinated by drought. Its reoccurrence has been the source of much anguish and hardship. Partly as a result, it has also been central in the creation of a uniquely Australian identity. Climate risk is held up as part of the national psyche, yet the fit has never been an easy one. One of the clearest descriptions of the difficulty we have in coming to terms with drought is contained in our language. Arthur (2003) notes that the most common language associated with drought is the same as that used with war. Droughts are declared; drought is something we must combat and battle. Drought grips, creeps, bites, and decimates the land and people who are drought smitten, desperate, and ruined. They are pictured on cracked earth raising their eyes to the merciless, pitiless blue skies waiting for relief from drought-breaking rains. Country that is free from drought is called safe country—safe from the 'dread enemy', as drought was called in 1906, or 'our biggest enemy', according to a headline in 1995. Arthur (2003) poses the question that, if we use the language of invading pirates for a natural recurring phenomenon like drought, then Australia is a place perennially disrupted and disjointed and the experience of Australia must always be one of disappointment and suffering. She argued that our language is set to a default country, which is narrow, green, hilly, and wet and where rivers run, lakes are full, and drought is an exception.

In an overview of literature of the Australian pastoral zone, Lynch (2010) noted the sense of alienation and the need to add green gardens to make a home in the bush. He draws our attention to the words of the American writer Wallace Stegner, who famously argued that before Americans could live comfortably and responsibly in the American West, they would need to 'get over the color green' (Stegner 1992, 54). Also writing about the American West, Limerick (2009) noted Stegner's evocative references to a 'new palette of gray and sage green and sulphur-yellow and buff and toned white and rust-red, a new flora and a new fauna, a new ecology'. She built on Stegner's point, arguing that in an arid environment 'green is the color of disturbed ground; it is the color of places where you have utilised water you have diverted from other places. Bright green, in other words, is the color of disturbance'. Getting over the colour green features in Dorothea MacKellar's patriotic poem, where the love of a sunburnt country is contrasted with 'The love of field and coppice/Of green and shaded lanes'—to which she responds, 'I know but cannot share it/My love is otherwise' (MacKellar 1985). Likewise, when Henry Lawson wrote 'The City Bushman' as a critique of his contemporary Banjo Patterson, he included the line, 'For you sought the greener patches and you travelled like a gent' (Lawson 1978b). The contrast of bright green on a brown landscape is clearest with irrigation. The notion that the Australian landscape should be green rather than brown was strongly held by Alfred Deakin, who despaired over this continent that could produce so much in one season but was bare in others. His solution of irrigation schemes is now seen to have incurred economic and environmental costs (Davidson 1969), though they were instrumental in regional development.

Attempts to Normalise Drought and the Changing Role of Government

In April 2011 the Australian Primary Industries Ministerial Council (PIMC 2011) listed as one of the guiding principles of drought policy reform 'the acknowledgement that drought is just one of a number of hardships that can adversely impact farmers'. Similar points were made by the Drought Policy Review Task Force report (1990) that took the important step of removing the natural disaster status of drought. This sentiment was repeated in the Drought Review Panel (2004) and the Productivity Commission review of drought policy (Australia. Productivity Commission 2009), which further emphasised that drought is a private risk for farm enterprises. These and other reports, parliamentary enquiries, commissions, and task forces that have argued for normalising drought have been supported by agricultural economists. For example, Anderson (1979, 148) despaired that 'the majority of Australian farmers seem to subscribe to the view that rainfall variability, in particular rainfall deficiency, is merely an unfortunate occasional abnormality of the environment; and that, when drought does occur, government assistance will aid survival.' Kraft and Piggot (1989) asked, 'Why single out drought'? From the field of geography, Heathcote (1988) argued that farmers should rely less on official relief and recognise drought for what it was—a constant occupational hazard in Australia.

It is the history of official drought declarations that most clearly demonstrates the reluctance to accept the reality of the arid and variable climate— a reality that it is in abnormal rather than normal seasons that Australia is green like England. In keeping with the treatment of drought as a natural disaster, the news media typically report each successive drought as unprecedented and unpredicted (West and Smith 1996). Simmons (1993) points out that in Queensland, some shires had been either partially or completely drought declared 70 percent of the time since 1964. In NSW, some districts had been drought declared for 3 months or more for 65 percent of the time and all of Victoria had been drought declared at least twice (Simmons 1993). After reviewing a series of definitions of drought, the Drought Policy Review Task Force (1990, 10–14) concluded that drought was relative, reflecting a situation whereby there was a mismatch between what they presented as agriculturists' expectations of a normal climate and the measured climate at that time.

A key driver of this focus on drought has been the fact that it poses particular challenges for agriculture due to the sector's dependence on and exposure to climate. Agriculture's reliance and influence on and vulnerability to nonhuman nature are key to the sector's unique status in Australia and elsewhere. Over recent decades, the special status of both drought and agriculture has been challenged within Australia by a series of policies that attempt to normalise both. These efforts to decrease the 'exceptionalism' of both agriculture and drought have been driven in part by a desire to decouple agricultural fortunes from climatic ones. This move stems not only from concern for those in farming, but also from broader social, economic, environmental and political motivations, reflecting the way agriculture has come to be viewed critically as a threat to public goods (for example, soil, biodiversity, water, government funding).

Given the public-good rationale for improving drought management in agriculture, government continues to play a role in the area. On questions of climate risk management more generally, the two main roles government plays are emergency management and provision of information. A key part of normalising drought in the Australian context has been declassifying it as a natural disaster. As such, drought is no longer seen, in theory at least, as a legitimate target for emergency management efforts and government funding in this area has consequently been cut. Partly to compensate for this reduced government role, partly to reinforce the process of normalising both drought and agriculture, and partly in keeping with its broader agenda to privatise risk (Quiggin 2009), government has instead privileged the role of information provision in its efforts to improve drought management in agriculture. More specifically, the form of information it has privileged in this effort is science. The special status attributed to science on the issue of drought management in agriculture reflects and reinforces the dominance of scientific perspectives on questions of both climate and agriculture in the past.

Drought throughout Overlapping Eras of Australian Farming

Hayman and McCarthy (in press) apply Richard Bawden's (1990) phases of Australian agriculture to the history of irrigation; here we apply and adapt them more broadly to an understanding of drought. According to Bawden, post-European Australian society first encouraged farmers to be pioneers and *'just produce'*. In this era, all and any agricultural produce was welcome, as were the associated efforts to claim and tame the new country. As the colony became established, government policy and industry leadership then tried to shift agriculture from small-scale survivalist mode to larger scale market mode to build the nation and supply resources to the empire. Farmers were encouraged to envisage themselves as major sources of production, gaining social kudos for their success as well or instead of the mere nobility of their vocation. The message during this era of professionalisation was *'produce more'*. More recently, such a focus on the quantum of production has been revitalised by calls to increase agricultural outputs to feed the world's increasingly sizeable population.

In addition to producing more, it soon became apparent that a more 'business style', export-oriented, and economically sustainable agriculture required a focus not only on how much was produced but also on how it was produced. Costs and profitability became key concerns. A new message was directed down to farmers: *'Produce more efficiently'*. Part of the rationale here was that farmers had to better manage their finances in order to cope with reduced levels of government protection as the sector was progressively deregulated in keeping with the government's deregulatory agenda.

Attempts since 1992 to reduce financial support related to drought are part of this broader economic rationalist push. Besides financial costs, environmental costs also began to cause concern as some long-term, long-distance consequences of industrial farming techniques began to manifest. This led to the final message Bawden notes: *'Produce more carefully'*.

Each of these requests has been additional. Still today, Australian farmers are seen as part pioneer, part entrepreneur, and part environmental custodian. These discourses have all had impacts on farmers' decision making and risk management. So too has the discourse that has emerged since Bawden (1990). Stimulated in part by the millennium drought and breaking rains, Australian agriculture is now also in a 'climate change era'; despite widespread denial (see Chapters 3 and 12) it is occurring. The message for farmers here is a reinforcement of each of the preceding messages, plus a call to *'produce more adaptively'* (or even *'produce where and when possible'*). This focus on adaptation rather than mitigation reflects the current status of the discussion in the agricultural sector and Australian society more broadly. Still highly controversial within Australian agriculture (WIDCORP 2009), the topic of climate change has (somewhat inadvertently) increased attention on climate variability and especially climatic extremes, including drought. It has also strengthened the role of science within the sector.

In Australia, the preceding history can be represented as three overlapping eras (Table 4.1). The first two represent the coarse distinction in Bawden's stages between a 'production only' focus and a 'production plus' focus, incorporating sustainability concerns; this is encapsulated in some overviews as productivist and multifunctional agriculture (see, for example, Cocklin and Dibden 2006; Holmes 2010; Argent 2011). The third is the climate change era as introduced before. These three eras are now discussed in turn in order to map the rise of different discourses about drought within Australian agriculture over time. More specifically, two or three framings of drought associated with each era are identified in order to highlight the ambiguous way drought is conceived in Australian agriculture and the way that it has revealed numerous underlying vulnerabilities and catalysed associated changes within the agricultural sector. The particular role of science in constructing these notions of risk and the ongoing dominance of the original production era are also highlighted.

Drought as a Production Risk: 1900 to Present

Drought as a Disruption

The early settlers realised that the rainfall in Australia was certainly much more variable than in England (London and Sydney have approximately the

TABLE 4.1

Overlapping Eras of Agricultural Development, How the Problem of Drought Is Defined in Each Era, and the Stylised Roles of Scientists and Farmers

Overlapping Eras	Problem Definition	Stylised Roles of Scientists and Farmers
Production and productivity, 1900 to present	Drought must be managed because it reduces farm production. Recently, this has been reinforced by renewed interest in feeding the growing global population.	Science provides material (varieties, breeds, irrigation) and information on climate risk and seasonal forecasts for optimum crop choice and input levels. Much of this information is 'packaged' in decision support systems. Farmers use techniques and information to optimise productivity within climate constraints.
Sustainability, 1980 to present	Drought must be managed because it presents a risk to the resource base—for example, wind erosion, pasture degradation, biodiversity loss, and weed invasion.	Science provides techniques (for example, chemicals and equipment for conservation farming; drip irrigation) and information for sustainable stocking rates, fertiliser inputs, and irrigation applications. Farmers use techniques and information to minimise land degradation.
Climate change, 2000 to present	Drought must be managed as part of farmers and rural communities adapting to a changing climate	Science provides information on future climates, the impact of the changes, and appropriate adaptation options. Farmers use information and techniques to adapt to worrisome but uncertain climate change projections.

same annual average rainfall; the difference is the variability.). As pointed out by Nicholls (2005), the settlement was greeted by an El Niño drought 3 years after settlement. He cites Captain Arthur Phillip as reporting in 1791,

> [S]o little rain has fallen that most of the runs of water in the different parts of the harbour have been dried up for several months and the run which supplies this settlement is greatly reduced. I do not think it is probable that so dry a season often occurs.

This early remark covered both the European despair over the climate and an optimistic sense that drought was an aberration or something rare. Nicholls has traced drought as a recurring theme through Australian history. He points out that the late 1800s, in both America and Australia, was a time of rural optimism based partly on the belief that rain followed the plough. However, this optimism was ill founded. In 1888, as the centennial celebrations commenced, the worst drought yet seen in the Colonies began. Henry Lawson wrote in the bulletin of December 1888: 'Beaten back in sad dejection/After years of weary toil/On the burning hot selection/where the drought has gorged his spoil' (Lawson 1978a).

While drought does explain much of Australian agricultural history, we need to be careful not to overlook other major challenges for European farmers in this colonial outpost such as clearing trees, the lack of draught animals, and a shortage of labour (Henzell 2007). Drought was not the only climatic issue facing the first colony. According to Henzell (2007), there were 10 high floods from the Hawkesbury River between 1799 and 1819, with regular reporting of drowning. Settlers faced a trade-off between fertile river flat soils and safety for themselves, families, crops, and livestock. Nevertheless, a feature of Australian farming, especially as it moved into the interior, was dealing with drought. The role of agricultural scientists in selecting crop varieties that are drought tolerant, mainly through flowering earlier in the season, is well documented (Fischer 2010) and can be dated back to William Farrer in the late 1800s to advances in functional genomics. An obvious application of science and technology to the disruption caused by drought was irrigation schemes. In the mid-1880s, Victorian Minister Alfred Deakin, who was responsible for water resources, held that if the state of Victoria was 'to progress...and use her abundant natural advantages and secure to the agricultural population of her arid districts a permanent prosperity, it would be through irrigation' (cited in Pownall 1968, 51–52).

Drought as a Business Risk

Keogh et al. (2011) drew attention to the extreme volatility of Australian agriculture as a reason for government programmes to address drought risk. When measured as variability of output, they showed that agriculture is the most volatile industry sector in the Australian economy—more than 2.5 times the average of all industries and significantly more volatile than the next ranking industries of insurance and construction, which were 1.5 times the average of all industries. With relatively low levels of government support, Australian farmers face a degree of price volatility that is as high as any other OECD country and the highest degree of production volatility (Kimura and Antón 2011).

Dillon (1992) defines risk as uncertainty with consequences. Climate uncertainty makes farming risky because the optimum crop type, crop area, input levels, and stocking rates differ from season to season depending on the climate, yet decisions need to be made and scarce resources allocated prior to knowledge of how the season will turn out. Climate science and farm management economics have information and procedures that can help with this decision making. The first level of information is access to long-term climate records (climatology) and the interpretation and transformation of this data to simulated crop and pasture yield. Since the late 1980s seasonal climate forecasts have been used with crop and pasture simulation models as guidance for decision making. These do not eliminate the risk, but rather partition the climatological distribution into conditional probability distributions with a mean, variance, and skew that differ from the base climatology. The value of these probability distributions lies in the fact that they enable the decision maker to better allocate resources.

Heathcote (1988), in his overview of drought management in Australian farming, was encouraged by the emerging science of climate forecasting based on sea surface temperatures in the Pacific to predict the coming season. Hammer and Nichols (1996) argue, 'We are confronted with unprecedented opportunities to tune our agricultural systems in a way that improves their sustainable land use. We have a seasonal forecasting capability. We have started to think through how we can best use the knowledge that the next season is not a total unknown'. In a report to the US academies of science, Easterling (1999, xi) claimed that seasonal climate forecasts based on the understanding of atmosphere and oceans were the premier advance of the atmospheric sciences in the twentieth century. Others have referred to seasonal climate forecasts based on El Niño Southern Oscillation (ENSO) as the meteorological equivalent to DNA (cited by Davis 2001) or the New Green Revolution (cited by Hansen 2002). Using the economic concepts deployed by some critics of older forms of drought assistance, the scientific information increases the knowledge of producers, therefore enabling them to make rational decisions. Consequently, those who are unviable should not be so because of a drought. As a relatively short-term stressor, drought, however, often reveals longer-term vulnerabilities that otherwise may have been ignored. This includes farm enterprises struggling with small scale (Malcolm 1994). The idea that some farmers' struggles under drought are primarily caused by longer-term factors, such as their failure to adapt to the economic rationalist agricultural policy trajectory, underpins criticism of government drought assistance to struggling farmers: the idea that those such assistance helps are already beyond help for other reasons (see, for example, Australia Productivity Commission 2009, 99).

Drought as a Risk to the Natural Resource Base: 1980s to Present

Drought as an Environmental Risk in Dryland Farming

As the role of farming has shifted from a predominant emphasis on just production to notions of sustainable production, the risk posed by drought has broadened to include natural resource management. McKeon et al. (2004)

review the eight major degradation episodes across Australia's grazed range-lands from the 1890s to the 1980s. In each case there was a run of good seasons during which stocking rates were increased, followed by a prolonged drought. In most cases there was irreversible damage to desirable species and in all cases there was irreversible soil loss, primarily through wind erosion. They argue that improved alert systems that combined resource monitoring (including remote sensing) and climate forecasting were important contributions from science to limit future degradation events. As acknowledged by the authors, a seasonal climate forecast for above-average rainfall that encouraged graziers to increase stocking rate could increase the risk if the season turned out to be much drier than forecast. Just as drought can expose and exacerbate underlying problems of farm viability, it can also reveal long-term trends in resource degradation. Stafford-Smith and McKeon (2007) drew attention to a core concept of resilience theory: distinguishing and treating underlying slow variables rather than the more obvious fast variables. The example they use is of a drought bankrupting farm families living on an eroded landscape with no stored capital. The drought is the fast variable, while the long-term degradation and nonviable farming are slow variables. The point is that treating drought will not solve the problem of past erosion events or farm viability.

As the science supporting agriculture has moved from production agriculture to farming systems to applied ecology, Australia's distinctiveness is often emphasised in scientific writing on Australian climate and ecology. Representing a new, albeit mild, form of nationalistic rhetoric, this writing is reminiscent of the earlier pride in Australia's unique climate and landscapes discussed previously. In their paper for the Prime Minister's Science and Engineering Council, for example, Clewett et al. (1995) contend that international research on the application of knowledge about climate variability to agriculture was rudimentary, but that Australia was at the forefront because of decades of research in risky environments. In a further example, Flannery (1994) dedicated his study of ecology and climate to Australians trying to forge a nation out of the chaos of colonial history. We see here how scientific research on Australia's environment gains special kudos from the uniqueness of its subject matter, including an ongoing pride in and fascination with the harshness of the climate.

Drought as an Environmental Risk in Irrigation

In his book *The Water Dreamers,* historian Cathcart (2009) reviews earlier Australians' efforts to try to solve the problem of aridity, with explorers setting off in hope of finding an inland sea and being disappointed by the dead heart. Cathcart's argument is that European Australia has grappled with water and aridity for all of its history. This was perhaps the clearest in the water crisis at the beginning of the twenty-first century. A major part of coming to terms with being a water-limited environment has been escalating questions about the role of irrigation. Drought in the Murray–Darling Basin has revealed the pre-existing vulnerability of administrative overallocation of water rights (Connell 2005) as well as weaknesses in existing scientific knowledge of the water system (for example, poor understanding of the relationship between surface and ground water)—all of which add scientific, social, and policy uncertainty to climate uncertainty for irrigators (Wei et al. 2011).

Irrigators have had to come to terms with governments no longer guaranteeing a high-security water supply, with annual allocations varying according to what is stored in dams and complex year-to-year scientific modelling of how much is likely to be available (Khan 2008; Wei et al. 2011; Wittwer and Griffith 2011). Irrigators now have to share what is available in dams at the start of the irrigation season and have to compete with stock and domestic use, urban users, and environmental requirements. Irrigators have to manage the risk of future allocations by carrying water over from the previous season, estimating the chance of water becoming available and trading temporary and permanent water rights. As discussed in more detail in Hayman and McCarthy (in press), the drought has also highlighted for irrigators some of the trade-offs between efficiency and resilience that ecologists have long known (Walker and Salt 2006). Highly efficient irrigation that eliminates losses such as drainage can lead to a buildup of salt in the root zone (Stevens 2002). Furthermore, the move to deficit irrigation and minimising water use in wine grape production through smaller canopies seems to have penalised some growers in recent heat waves (Hayman and McCarthy 2009; Webb and Watt 2009).

Drought as a Climate Change Risk: 2000 to Present

The final 'era' of agriculture to discuss is climate change. Besides the need to understand climate change specifically, the discovery of anthropogenic climate change is seen to have revealed further 'deficits' in the agricultural sector—farmers' 'climate literacy' and 'carbon literacy'—that have invigorated calls for more scientific research, development, and extension. Climate change has also complicated discussions about drought by reinforcing contrasting discourses about the phenomenon. At a coarse level, this has occurred simply as a result of climate change encouraging a continuous (rather than sporadic) focus on climate and the risks it poses. More specifically, the notion that drought is a persistent 'normal' threat has been strengthened, both by this unwavering interest in climate and by the central role that dryness has in projections and discussions about climate change in Australia. Climate change has also inadvertently helped to further normalise drought by encouraging climate sceptics (including the majority of

Australian farmers as discussed in Chapter 12) to emphasise the long and dominant role of climate variability in Australia's (farming) history as part of their rejection of the reality of anthropogenic climate change (Rickards forthcoming). In this way, farmers claim a special knowledge of and reinforce the narrative about the natural unpredictability of Australia's climate.

At the same time as climate change has helped to normalise drought, the novelty of the phenomenon of anthropogenic climate change means that all climatic phenomena, including drought, are now automatically 'abnormalised'. Theoretically if not yet empirically, climate is suddenly different from what we have known before on account of now being partly human induced. In Australia, this includes projected alterations to the timing, extent, intensity, climatic context, temperature, and predictability of droughts, thus reinforcing the tendency discussed earlier to claim that each drought is 'unprecedented'. As the stationarity of climate—'the idea that natural systems fluctuate within an unchanging envelope of variability' (Milly et al. 2008, 573)—gives way to the nonstationarity of human-influenced climate, the idea that each successive drought is 'unprecedented' now has a new level of meaning.

Besides reinforcing pre-existing discourses about drought, climate change has also introduced new ones. In part, this reflects the different ways in which climate extremes in general have been framed by researchers in their attempts to unravel the difficult relationship between natural climate variability and anthropogenic climate change. New interpretations of drought also stem from characteristics unique to drought, including that most recently experienced in southeast Australia. Four new framings especially pertinent to Australian agriculture are now discussed:

- Drought as an analogue and source of adaptive capacity
- Drought as an obstacle to adaptation
- Drought as a window of opportunity
- Drought as a signal event

Drought as an Analogue and Source of Adaptive Capacity

In trying to understand how society can best adapt to climate change, adaptation researchers have been looking for analogues of the sorts of conditions that climate change is projected to bring about (Ford and Keskitalo 2010). In Australia, drought qualifies as a temporal analogue on two counts. Droughts are predicted to be more common and acute (Hennessy et al. 2008), and the background trend is toward more drought-like conditions (hotter and drier) (Hennessy and Whetton 2010). While limited by the way in which droughts in the future may differ substantially from those experienced to date (discussed earlier), the idea that droughts are a 'window onto the future' for Australia has further raised the profile of the phenomenon.

One result of the expected overlap between past drought conditions and projected future conditions is a sense of familiarity and thus confidence about what climate change holds. In this way, drought has also been framed as a valuable teacher and consequent origin of adaptive capacity (the ability and resources to adapt to climate change). Such a positive interpretation of adaptive capacity is especially apparent among the farming community. Claims about farmers' unique experience with drought—and thus future climates—now add to discourses about farmers' special abilities and status. As Ash et al. (2000, 269) write, for example, 'Australia's primary industries, which already have to manage extreme climatic variability, are in a position to lead Australian industry in innovative and proactive responses to climate change'. This positive interpretation of the relevance of farmers' past drought experiences to future climate demands depends on either a view of future climate as predominantly similar in kind to past climate or as having instilled in farmers a general capacity to manage future climates no matter what form they take. This capacity has been assisted by the many locally situated scientific research projects on farming issues that have been conducted (usually with farmer input) throughout Australia. Combined with an emphasis on the 'upside' risks that climate change may create, a sense that the future is deeply uncertain has been embraced by some in the industry with a sense of excitement, revitalising a past image of farmers as 'pioneers' courageously setting off into the unknown (see, for example, Anderson 2010). Whether the future is broadly similar to or profoundly different from the past, this view of drought tends to be based on a favourable assessment of the success of farmers' risk management and learning to date, and on the ongoing relevance of local research and development efforts. As we now discuss, however, that view is also contested.

Drought as an Obstacle to Adaptation

Like other climate extremes, drought is also framed as an obstacle to successful climate change adaptation. This stems firstly from the fact that drought and related coping strategies (for example, working harder, taking on debt, reducing inputs) often erode farm families' business and personal reserves, reducing their capacity to make longer term adaptations to climate change (Rickards forthcoming). While sometimes framed as an automatic and easy process, successful adaptation to climate change can involve substantial risks and costs that farm families are ill equipped to take on when already occupied with surviving drought's effects. Drought in this sense is also framed as a 'distraction' from the more strategic change desired of farmers. Associated with the criticisms of farming inherent to the production and sustainability eras discussed previously, drought is further viewed as an obstacle to successful climate change adaptation to the extent that it is a source of maladaptive rather than adaptive learning. In counterpoint to the idea that farmers are well equipped to deal with future drought-like conditions because they have dealt with droughts in the past, this view argues that farmers' past drought management efforts have been unsuccessful in significant ways, resulting in substantial 'adaptation deficits' for them and others (Burton 2011) (for example, soil degradation, relationship stress, expectations of assistance). By reinforcing bad habits and reducing reserves, drought is seen to distort farmers' approaches and obstruct their capacity to manage future climate successfully.

Drought as a Window of Opportunity for Transformational Change

A more optimistic view of drought-induced crises is provided by an interpretation of drought as a welcome window of opportunity for positive (as opposed to negative) transformational change. Reflecting the idea of 'trigger points' and 'thresholds' in resilience thinking (Walker and Salt 2006), this view sees drought and other crises as a valuable crack in the foundations of an established system, forcing those involved to re-evaluate radically what they have been doing and create a new improved system. Prevalent in discussions of climate change adaptation, this view of climatic extremes as useful is often premised on the assumption that those involved have enough capacity to envisage and enact such positive change (Birkmann and Buckle 2010). What constitutes positive change is, of course, contestable and costs and benefits of transformational change processes may not be evenly distributed across the different parties involved. The positive side of transformation may only be apparent to those operating at a higher scale of governance. Research on 'disaster politics' suggests that governments often seize disasters as valuable periods of increased political legitimacy for taking decisive leadership, allowing substantial changes to be ushered in (Pelling and Dill 2010).

Disaster itself can have what those at a high level of organisation perceive to be a welcome restructuring effect, eliminating elements of a system that they believe need to be eliminated. This survival of the fittest mentality is evident in the idea that drought impacts are 'natural' (not a reflection of pre-existing policies) and therefore 'good'. In Australia, criticisms of the National Drought Policy, for example, are based on the argument that drought (in concert with other unquestioned contextual factors) accelerates a desirable restructuring process within the farming sector, forcing unviable or poorly drought-adapted farm families out and freeing up land for more efficient and/or more environmentally sustainable farmers to purchase. This view implies that drought impacts distribute a form of natural justice. It represents drought risk as a risk created by human action, rather than a risk created by accident or fate in the way that most natural disasters are viewed (Li 2010). Notably, however, the role of human agency in (inadvertently) shaping the climate and thus the occurrence of drought is not prominent in these discussions. That is, human agency is limited to discussion of drought impact, not drought causation. As we now discuss, however, drought has prompted people to face the topic of human-induced climate change.

Drought as a Signal Event

Related to the framings of drought as an analogue and window of opportunity for change, the final framing of drought is of it as a catalyst for concern about human-induced climate change. This framing of drought as a 'signal event' is evident in discussions and actions surrounding the Big Dry or millennium drought across south-eastern Australia, a prolonged dry spell from, in some areas, 1997 to 2009 (Vernon-Kidd and Kiem 2009; Timbal et al. 2010). A 'signal event' refers to an event that is symbolically charged and becomes difficult to separate from future discourse on the matter. For example, any discussion of nuclear safety after 1986 is likely to include a reference to Chernobyl. Wilkins (2000) applied the term to the 1997 El Niño event in the United States, which was the strongest El Niño on record and had a major impact on US public perceptions. Key to the longevity of media interest in the event was the ability of climate science to provide an explanatory narrative as a background to the spectacular pictures of its impacts, underlining the interconnectedness of scientific and lay discourses about events such as drought (Li 2010). In particular, interest was sustained by debate about whether it was caused or enhanced by human-induced climate change and/ or was a window on a future under climate change (an analogue, as discussed before).

In Australia, the Big Dry has served as a signal event in similar ways. First of all, it has numerous 'unprecedented' and thus climate change-like characteristics, including being hotter (Nicholls 2004) and involving especially low runoff (CSIRO and Bureau of Meteorology 2010). Facilitated by a new level of public interest in climate science—including a perverse excitement when climate records are broken—these novel elements of the drought have served to reinforce the message that the climate is shifting. At the same time, the extended length of the drought has normalised the experience of living with drought-like conditions, reinforcing the other message that drought is an unremarkable part of Australian life. The Big Dry has also reinvigorated interest in farming and rural Australia. Some of this stems from concern about farm families' and rural communities' well-being, with a wave of social research being initiated into the acute difficulties some rural people are experiencing (for example, Anderson 2008; Kenny et al. 2008; Caldwell and Boyd 2009; Edwards and Gray 2009; Kiem and Askew 2010). In particular, there has been a widespread campaign on mental health in rural communities (Brumby and Willder 2009), focussed especially on farmers and the perceived limitations of their coping strategies (including the fact that they are, ironically, too 'self-reliant') (Alston and Kent 2008; Berry and Hogan 2011). In this way, aspects of farmers' psychology have been 'revealed' by the drought as another apparent vulnerability in the sector.

Some of the interest in farming catalysed by the Big Dry has been more overtly critical. As discussed earlier, this includes the latest review of the National Drought Policy, prompted by concerns that the 'wrong farmers' have been being supported by the policy to date and that drought risk needs to be more fully transferred to farm businesses to help determine which ones are ultimately viable. It also includes interest in agriculture's role as a cause of human-induced climate change. A large new area of scientific research has emerged to investigate the greenhouse intensity of different agricultural practices, as well as the potential for agriculture to sequester carbon. Critical interest in agriculture induced by the Big Dry also includes strong concerns about irrigation water use. The Big Dry has been called the first irrigation drought (Horticulture Australia Council 2008) or 'double drought' (Rickards and Tucker 2009), with not only local rainfall but also distant water supplies failing.

Conventional drought images of desiccated crops and starving stock have been complemented by pictures of empty government-run dams and abandoned vineyards. These pictures underline not only the geographic extent of the recent drought, but also the limitation of relying on irrigation as a drought adaptation, including the 'high-security water' on which permanent horticulture historically has been based. Pictures of dying river gums have further shifted attention from the issue of water supply to water demand, pointing to the difficult question of how any available water is allocated between agriculture and environmental users such as Murray–Darling ecosystems. Resultant political and institutional uncertainty about water allocations has added to the risks irrigators have had to face during the drought. Consequent changes made to the allocation system (such as a greater degree of flexibility in allocations) are likely to have a long-lasting effect. More broadly, debates about the merits of irrigated agriculture have included discussion of the greenhouse costs of mechanised irrigation and added to calls for climate change adaptation to encompass changes not only in agricultural land management but also to land use, notably shifts out of agriculture. In these ways, the Big Dry has catalysed a substantial shift in not only the physical but also the social and political climate in which agriculture exists.

Conclusion

According to Brown (2010), wicked problems are multicausal and develop through time. This often leads to conflicting goals, which makes them unstable. The instability means that problem solvers (in this case, farmers, scientists, and policy makers) are dealing with a moving target. In this chapter we have examined how drought, farming, science, and climate change have interacted over time. It follows that this interaction will continue. Our argument has been that these privileged topics have acted to enhance each other (Figure 4.1). As identified by Botterill (2009) and others, drought as a special risk relies heavily on agricultural exceptionalism. The emerging issue of climate change can be seen as both reinforcing and destabilising this relationship between drought and agricultural exceptionalism. The reinforcing component is a sense that agriculture has a role in food security in a climate-challenged world, which is reminiscent of the pioneering role of agriculture. Climate change has simultaneously raised new questions about appropriate use of land and water resources in Australia. How the established nexus of science and agriculture responds to the new relationship between drought and broader climate is yet to be seen. No doubt the relationships between all will be stretched and reshaped. It is likely, however, that all four will remain prominent.

References

Aitkin, D. 1985. 'Countrymindedness'—The spread of an idea. *Australian Cultural History* 4:34–41.

Alston, M., and J. Kent. 2008. The Big Dry: The link between rural masculinities and poor health outcomes for farming men. *Journal of Sociology* 44 (2): 133–147.

Anderson, D. 2008. Drought, endurance and 'the way things were': The lived experience of climate and climate change in the Mallee. *Australasian Humanities Review* 45:67–81.

———. 2010. Drought, endurance and climate change 'pioneers': Lived experience in the production of rural environmental knowledge. *Cultural Studies Review* 16 (1): 82–101.

Anderson, J. R. 1979. Impacts of climatic variability in Australian agriculture: A review. *Review of Marketing and Agricultural Economics* 47:147–177.

Argent, N. 2011. Trouble in paradise? Governing Australia's multifunctional rural landscapes. *Australian Geographer* 42 (2): 183–205.

Arthur, J. M. 2003. *The default country: A lexical cartography of twentieth-century Australia.* Sydney: UNSW Press.

Ash, A., P. O'Reagain, G. McKeon, and M. Stafford Smith. 2000. Managing climate variability in grazing enterprises: A case study of Dalrymple shire, north-eastern Australia. In *Applications of seasonal climate forecasting in agricultural and natural ecosystems: The Australian experience,* ed. G. L. Hammer, N. Nicholls, and C. Mitchell, 253–270. Dordrecht: Kluwer Academic.

Australia. Productivity Commission. 2009. Government drought support. Report No 46. Melbourne.

Bawden, R., ed. 1990. Towards action researching systems. In *Action research for change and development,* ed. O. Zuber-Skerritt, Brisbane: CALT Griffith University.

Berry, H. L., and A. Hogan. 2011. Climate change and farmers' mental health: Risks and responses. *Asia-Pacific Journal of Public Health* 23 (2 suppl): 119S–132S.

Birkmann, J., and P. Buckle. 2010. Extreme events and disasters: A window of opportunity for change? Analysis of organizational, institutional and political changes, formal and informal responses after mega-disasters. *Natural Hazards* 55 (3): 637–655.

Botterill, L. C. 2007. Responding to farm poverty in Australia. *Australian Journal of Political Science* 42 (1): 33–46.

———. 2009. The role of agrarian sentiment in Australian rural policy. In *Tracking rural change: Community, policy and technology in Australia, New Zealand and Europe*, ed. F. Merlan and D. Raftery, 59–78. Canberra: ANU ePress.

Brown, V. A. 2010. Conducting an imaginative transdisciplinary inquiry. In *Tackling wicked problems through the transdisciplinary imagination*, ed. V. A. Brown, J. A. Harris, and J. Y. Russell, 103–114. London: Earthscan.

Brown, V. A., and P. M. Deane. 2010. Towards a just and sustainable future. In *Tackling wicked problems through the transdisciplinary imagination*, ed. V. A. Brown, J. A. Harris, and J. Y. Russell, 3–15. London: Earthscan.

Brumby, S., and S. Willder. 2009. The sustainable farm families project: Changing attitudes to health. *Rural and Remote Health: The International Journal of Rural and Remote Health Research, Education, Practice and Policy* 9:1012.

Burton, I. 2011. Climate change and the adaptation deficit. In *The Earthscan reader on adaptation to climate change*, ed. E. Schipper and I. Burton, 89–98. London: Earthscan.

Caldwell, K., and C. P. Boyd. 2009. Coping and resilience in farming families affected by drought. *Rural and Remote Health: The International Journal of Rural and Remote Health Research, Education, Practice and Policy* 9:10.

Cathcart, M. 2009. *The water dreamers: The remarkable history of our dry continent*. Melbourne: Text Publishing.

Clewett, J. F., W. Kininmonth, and B. J. White. 1995. A framework for improving management of climatic risks and opportunities. In *Sustaining the agricultural resource base, 12th meeting of Prime Minister's Science and Engineering Council*. Canberra, 48–60.

Cocklin, C., and J. Dibden. 2006. From market to multifunctionality? Land stewardship in Australia. *Geographical Journal* 172 (3): 197–205.

Connell, D. 2005. Managing climate for the Murray–Darling Basin 1850–2050. In *A change in the weather: Climate and culture in Australia*, ed. T. Sherratt, T. Griffiths, and L. Robin. Canberra: National Museum of Australian Press.

CSIRO and Bureau of Meteorology. 2010. State of the climate 2010. Available from www.bom.gov.au/inside/eiab/State-of-climate-2010-updated.pdf

Davidson, B. 1969. *Australia: Wet or dry?* Melbourne: Melbourne University Press.

Davis, M. 2001. *Late Victorian holocausts: El Nino famines and the making of the Third World*. New York: Taylor & Francis.

Dillon, J. L. 1992. The farm as a purposeful system, Miscellaneous publication no 10. Armidale: Department of Agricultural Economics and Business Management, University of New England

Douglas, M., and A. Wildavsky. 1982. *Risk and culture. An essay on the selection of technical and environmental dangers*. Berkeley: University of California Press.

Drought Policy Review Task Force. 1990. *National drought policy*. Drought Policy Review Task Force. Canberra: Commonwealth of Australia.

Drought Review Panel. 2004. *Consultations on national drought policy—Preparing for the future*. Canberra: Department of Agricultural, Fisheries and Forestry. www.daff.gov.au/droughtassist

Easterling, W. E., ed. 1999. *Making climate forecasts matter*. Washington, DC: National Academy Press.

Economist. The century's defining issue 2009. Available from http://www.economist. com/blogs/freeexchange/2009/12/the_centurys_defining_issue

Edwards, B., and M. Gray. 2009. A sunburnt country: The economic and financial impact of drought on rural and regional families in Australia in an era of climate change. *Australian Journal of Labour Economics* 12 (1): 23.

Fischer, R. A. 2010. Farming systems of Australia: Exploiting the synergy between genetic improvement and agronomy. In *Crop physiology. Applications for genetic improvement and agronomy*, ed. V. O. Sadras and D. F. Calderini, 23–54. New York: Elsevier.

Flannery, T. F. 1994. *The future eaters, an ecological history of the Australasian lands and people*. Sydney: Reed Books.

Ford, J. D., and E. C. H. Keskitalo. 2010. Case study and analogue methodologies in climate change vulnerability research. *WIREs Climate Change* 1:19.

Garnaut, R. 2008. *The Garnaut climate change review*. Report to the Australian Government. Melbourne.

Hammer, G. L., and N. Nicholls. 1996. Managing for climate variability—The role of seasonal climate forecasting in improving agricultural systems. *2nd Australian Conference on Agricultural Meteorology*, at Brisbane.

Hansen, J. W. 2002. Realizing the potential benefits of climate prediction to agriculture: Issues, approaches, challenges. *Agricultural Systems* 74:309–330.

———. 2010. Obama's second chance on the predominant moral issue of this century. Available from http://www.huffingtonpost.com/dr-james-hansen/obamas-second-chance-on-c_b_525567.html

Hayman, P. T., and M. McCarthy. 2009. Assessing and managing the risk of heat waves in South Eastern Australian winegrowing regions. *The Australian and New Zealand Grapegrower and Winemaker* 542:45–49.

———. In press. Irrigation and drought in a southern Australian climate that is arid, variable and changing. In *Drought in arid and semi-arid regions: A multi-disciplinary and cross-country perspective*, ed. K. Scwabb, J. Albiac, J. Connor, R. Hassan, and L. Meza-Gonzalez. Dordrecht: Springer Publishing.

Head, B. 2008. Wicked problems in public policy. *Public Policy* 3 (2): 101–118.

Heathcote, R. L. 1973. Drought perception. In *The environmental, economic and social significance of drought*, ed. J. V. Lovett, 17–40. Sydney: Angus and Robertson.

———. 1988. Drought in Australia: Still a problem of perception? *Geojournal* 16:347–397.

Hennessy, K., R. Fawcett, D. Kirono, F. Mpelasoka, D. Jones, J. Bathols, P. Whetton, M. Stafford Smith, M. Howden, C. Mitchell, and N. Plummer. 2008. *An assessment of the impact of climate change on the nature and frequency of exceptional climatic events*. Canberra: Bureau of Meteorology and CSIRO.

Hennessy, K., and P. Whetton. 2010. Climate projections. In *Adapting agriculture to climate change: Preparing Australian agriculture, forestry and fisheries for the future*, ed. C. J. Stokes and M. Howden, 13–20. Melbourne: CSIRO Publishing.

Henzell, T. 2007. *Australian agriculture: Its history and challenges*. Melbourne: CSIRO.

Hepburn, J. Climate denial, science and genetic engineering 2011. Available from http://blogs.crikey.com.au/rooted/2011/08/02/genetic-engineering-science-and-climate-denial/

Holmes, J. 2010. Divergent regional trajectories in Australia's tropical savannas: Indicators of a multifunctional rural transition. *Geographical Research* 48 (4): 342–358.

Horticulture Australia Council. 2008. Submission to drought review: Response to the Productivity Commission's inquiry into drought support policy. Submission 66, December.

Ison, R. 2010. *Systems practice: How to act in a climate change world.* London: Springer and The Open University.

Kenny, P., S. Knight, M. Peters, D. Stehlik, B. Wakelin, S. West, and L. Young. 2008. *It's about people: Changing perspectives on dryness—A report to government by an expert social panel.* Canberra: Commonwealth of Australia.

Keogh, M., R. Granger, and S. Middleton. 2011. *Drought Pilot Review Panel: A review of the pilot of drought reform measures in Western Australia.* Canberra: Commonwealth of Australia.

Khan, S. 2008. Managing climate risks in Australia: Options for water policy and irrigation management. *Australian Journal of Experimental Agriculture* 48 (3):265–273.

Kiem, A. S., and L. E. Askew. 2010. *Drought and the future of small inland towns: Drought impacts and adaptation in regional Victoria, Australia. A report for the National Climate Change Adaptation Research Facility (NCCARF) Synthesis and Integrative Research Program: Historical case studies.* Newcastle, Australia.

Kimura, S., and J. Antón. 2011. Risk management in agriculture in Australia. OECD Food, Agriculture and Fisheries Working Papers, No 39: OECD Publishing. http://dx.doi.org/10.1787/5kgj0d8bj3d1-en.

Kraft, D., and R. Piggott. 1989. Why single out drought? *Search* 6:189–192.

Lawson, Henry. 1978a. Beaten back. In *Henry Lawson favourite verse,* ed. N. Keesing, 97. Melbourne: Nelson.

———. 1978b. The city bushman. In *Henry Lawson favourite verse,* ed. N. Keesing, 152. Melbourne: Nelson.

Li, G. 2010. Calculating community risk: A transdisciplinary inquiry into contemporary understandings of risk. In *Tackling wicked problems through the transdisciplinary imagination,* ed. V. A. Brown, J. A. Harris, and J. Y. Russell, 171–179. London: Earthscan.

Limerick, P. 2009. Celebrating Wallace Stegner's most quotable words. PERC Report Property and Environment Research Centre.

Lynch, T. 2010. Literature in the arid zone. In *The littoral zone. Australian contexts and their writers,* ed. C. A. Cranston and R. Zeller, 72–92. Amsterdam: Rodopi.

MacKellar, D. 1985. My country. In *My country: Australian poetry and short stories—Two hundred years,* ed. L Kramer, 472–473. Sydney: URE Smith Press.

Malcolm, L. R. 1994. Managing farm risk: There may be less to it than is made of it. *Proceedings of Conference: Risk Management in Australian Agriculture,* Armidale.

McKeon, G. M., W. B. Hall, B. H. Henry, G. S. Stone, and I. W. Watson. 2004. *Pasture degradation and recovery in Australia's rangelands. Learning from history.* Brisbane: Queensland Department of Natural Resources, Mines and Energy.

Milly, P. C. D., J. Betancourt, M. Falkenmark, R. M. Hirsch, Z. W. Kundzewicz, D. P. Lettenmaier, and R. J. Stouffer. 2008. Stationarity is dead: Whither water management? *Science* 319 (5863):573–574.

Nicholls, N. 2004. The changing nature of Australian droughts. *Climatic Change* 63:323–336.

———. 2005. Climatic outlooks: From revolutionary science to orthodoxy. In *A change in the weather,* ed. T. Sherratt, T. Griffiths, and L. Robin, 18–298. Canberra: National Museum of Australia Press.

Pelling, M., and K. Dill. 2010. Disaster politics: Tipping points for change in the adaptation of sociopolitical regimes. *Progress in Human Geography* 34 (1):21–37.

Primary Industries Ministerial Council. 2011. *Primary Industries Ministerial Council Communiqué*. PIMC 19, 15 April 2011.

Pownall, E. 1968. *The thirsty land: Harnessing Australia's water resources*. London: Coward-McCann.

Quiggin, J. 2009. *Risk shifts in Australia: The implications of the global financial crisis*. Risk and Sustainable Management Group, School of Economics and Political Science, University of Queensland, Australia.

Rickards, L. Forthcoming. *Critical breaking point? The effects of climatic and other stressors on farming families. A follow-up study*. Birchip Cropping Group.

Rickards, L., and K. Tucker. 2009. Challenges for Australian agriculture. In *Climate change: On for young and old*, ed. H. Sykes. Melbourne, 84–101. Future Leaders.

Simmons, P. 1993. Recent developments in Commonwealth drought policy. *Review of Marketing and Agricultural Economics* 61:101–138.

Stafford-Smith, D. M., and G. M. McKeon. 2007. Learning from episodes of degradation and recovery in variable Australian rangelands. *Proceedings of the National Academy of Sciences of the United States of America*.

Stegner, W. 1992. Thoughts in a dry land. In *Where the bluebird sings to the lemonade springs: Living and writing in the West*, 45–56. New York: Random House.

Stevens, R. M. 2002. Interaction between irrigation, salinity, leaching efficiency, salinity tolerance and sustainability. *Australian New Zealand Grapegrower and Winemaker* 466:71–76.

Timbal, B., J. Arblaster, K. Braganza, E. Ferabdez, H. Hendon, B. Murphy, M. Raupach, C. Rakich, I. Smith, K. Whan, and M. Wheeler. 2010. *Understanding the anthropogenic nature of the observed rainfall decline across south-eastern Australia*. CAWCR Technical Report No. 026. Melbourne: Centre for Australian Weather and Climate Research.

Turnpenny, J., and I. Lorenzoni. 2009. Noisy and definitely not normal: Responding to wicked issues in the environment, energy and health. *Environmental Science & Policy* 12 (3): 347–358.

Vernon-Kidd, D. C., and A. S. Kiem. 2009. Nature and cause of protracted droughts in southeast Australia: Comparison between the Federation, WWII and Big Dry droughts. *Geophysical Research Letters* 36:L22707.

Walker, B., and D. Salt. 2006. *Resilience thinking: Sustaining ecosystems and people in a changing world*. New York: Island Press.

Watts, M. 2003. Reporting unemployment. Winners and losers. George Munster Forum. Newcastle, Australia: Australian Centre for Independent Journalism.

Webb, L., and A. Watt. 2009. *Extreme heat: Managing grapevine response*. GWRDC and University of Melbourne.

Wei, Y., J. Langford, I. R. Willett, S. Barlow, and C. Lyle. 2011. Is irrigated agriculture in the Murray–Darling Basin well prepared to deal with reductions in water availability? *Global Environmental Change* 21:906–916.

West, B., and P. Smith. 1996. Drought, discourse, and Durkheim: A research note. *Journal of Sociology* 32 (1): 93–102.

WIDCORP. 2009. *Understanding farmer knowledge and attitudes to climate change, climate variability, and greenhouse gas emissions*. Prepared for the Department of Primary Industries, Victoria, Water in Drylands Collaborative Research Program. Melbourne.

Wilkins, L. 2000. Was El Nino a weather metaphor—A signal for global warming? In *The climate event of the century*, ed. S. A. Chagnon, 49–67. Oxford: Oxford University Press.

Wittwer, G., and M. Griffith. 2011. Modelling drought and recovery in the southern Murray–Darling Basin. *Australian Journal of Agricultural and Resource Economics* 55 (3): 342–359.

Section II

Science, Evidence, and Policy

5

Scientists and Drought Policy: A US Insider's Perspective

Gene Whitney

CONTENTS

Sound science is critical to good policy. Policy makers in the United States, including the President, Congress, state governors, and agency heads at all levels, must consider a variety of priorities, pressures, and inputs when making a decision or establishing a policy. No important issue can be distilled into a single societal discipline. Virtually all policies have economic, scientific and technical, legal, social, regulatory, moral, and political aspects, and the weight given to each aspect depends on the specific characteristics of the policy agenda and the policy makers. As global society becomes increasingly technically sophisticated, science is clearly one of the inputs needed to make good decisions in this technical policy milieu. Issues of energy, agriculture, space, medicine, telecommunications, national security, and climate are becoming more complex. For example, both medical and agricultural policies must now consider the science of genomics and all of the possibilities and potential liabilities associated with the ability of scientists to manipulate the genomic material of humans and other organisms. Likewise, policies related to disaster mitigation, preparedness, recovery, and resilience must increasingly consider aspects of climate change and the human and natural biological processes of adaptation. Drought policy lies at the nexus of disaster policy, agricultural policy, climate change policy, and water policy, and has become a 'wicked problem' in the United States.

Drought policy is difficult to formulate for a variety of reasons (Folger, Cody, and Carter 2010):

- Part of the difficulty is related to the definition of drought (see Chapter 1) and the characteristics of drought and the resulting impacts. Compared to other natural hazards and their consequential disasters, drought often emerges slowly and has therefore been characterised as a 'slow-onset disaster'; that is, the effects of drought grow slowly over time relative to sudden, rapid-onset disasters such as an earthquakes, hurricanes, or tsunamis.

- The scope of drought may be quite localised or may affect a whole continent, thus drawing in various levels of government and different economic sectors depending upon the specific conditions of the drought.

- It is not always obvious when a drought ends. A small amount of rain after a prolonged drought does not necessarily mean the event is over, and certainly not its consequences.

- Although drought affects a number of sectors that use water, a drought cannot easily be mitigated simply by hardening infrastructure, modifying building codes, or other concrete actions often associated with mitigating the impacts of other kinds of disasters. Drought mitigation might involve long-term changes in methods and behaviour.

- Measuring the economic impact of drought is sometimes difficult because estimates of lost crop value, lost recreational income, lost power generation, lost fisheries or wildlife, loss of water transportation, or the economic impact of water shortage on a particular community must be derived from estimates of what might have happened in the absence of the drought. There are no destroyed buildings to reconstruct, no burst dams to rebuild, and no sunken ships to replace after a drought. The return of precipitation after a drought restores agriculture for the next crop season, refills the streams and reservoirs, and provides the water resources for communities and industries. Once the rains return, the drought becomes history and life goes on.

Developing public policy to address the long-term preparedness, mitigation, and impacts of drought is difficult in a political culture that increasingly operates with a short-term perspective.

In the United States, politicians tend to focus on short-term problems because policies are promulgated by elected officials who operate on 2-year (House of Representatives), 4-year (President and state governors), or 6-year (Senate) re-election cycles. Preoccupation with re-election drives much of the policy process today. It is easier to gain political support and acceptance for funding recovery from an event than funding long-term investments in prevention or mitigation of impacts of an event that is uncertain and may vary in magnitude and impact. In other words, even though recovery may be much more expensive than prevention, it is often politically simpler to address an incident after it occurs than it is to prevent an incident or reduce its impact. These attitudes pervade US policy making in a number of arenas and have been exhibited graphically in drought policy.

The purpose of this chapter is to provide some perspective on the process of making drought policy in the United States and how scientists may inform or influence that process within the US policy-making framework.

Recent Drought Policy in the United States

Some of the key policy developments over the last few years are summarised here to provide context for the subsequent discussion of the policy-making process. Since 2005 (109th Congress), 328 pieces of legislation have contained the word *drought*, yet no major drought policy legislation has been passed.*

Congress passed into law the National Drought Policy Act in 1998 (Public Law 105-199). That law recognised that the United States needs drought policy, but rather than establishing such a policy, the Act created the National Drought Policy Commission to conduct a study to review existing drought policies (also see Chapter 10), to analyse the roles and responsibilities of various drought-related agencies and councils, and to come back to Congress with recommendations on what a national policy should contain.

The National Drought Policy Commission (the Commission) completed its work and published its final report in 2000 before being terminated, as stipulated in the law. The Commission report was 47 pages of excellent observations and recommendations (National Drought Policy Commission 2000). Among the findings of the Commission, the report issued the following policy statement:

> The Commission believes that national drought policy should use the resources of the federal government to support but not supplant nor interfere with state, tribal, regional, local, and individual efforts to reduce drought impacts. The guiding principles of national drought policy should be:
>
> 1. Favor preparedness over insurance, insurance over relief, and incentives over regulation.

---

* Based on a word search for 'drought' on legislation for multiple Congresses (as of September 27, 2011) using the Library of Congress Thomas online legislative search tool: http://thomas. loc.gov/home/thomas.php. The list includes mostly relevant references to natural drought, but also includes, for example, a reference to '49-year-long championship drought' of the Chicago Blackhawks hockey team.

2. Set research priorities based on the potential of the research results to reduce drought impacts.
3. Coordinate the delivery of federal services through cooperation and collaboration with nonfederal entities.

There is a recognition here of the realities of US federalism, a clear intention, as in recent Australian policy, to reduce the reliance on reactive policies (see Chapter 1), and a nod to evidence-based policy (Chapters 7 and 9).

In addition, the Commission provided the following set of recommendations and goals:

> We recommend first that Congress pass a National Drought Preparedness Act to establish a non-federal/federal partnership through a National Drought Council...The primary function of the Council is to ensure that the goals of national drought policy are achieved. Our five goals are:
>
> 1. Incorporate planning, implementation of plans and proactive mitigation measures, risk management, resource stewardship, environmental considerations, and public education as the key elements of effective national drought policy.
> 2. Improve collaboration among scientists and managers to enhance the effectiveness of observation networks, monitoring, prediction, information delivery, and applied research and to foster public understanding of and preparedness for drought.
> 3. Develop and incorporate comprehensive insurance and financial strategies into drought preparedness plans.
> 4. Maintain a safety net of emergency relief that emphasizes sound stewardship of natural resources and self-help.
> 5. Coordinate drought programs and response effectively, efficiently, and in a customer-oriented manner.

Subsequently, legislation was introduced to implement some of the findings and recommendations of the Commission. Most recently, 37 cosponsors introduced the National Drought Preparedness Act of 2005,* which had a companion bill in the Senate sponsored by 15 cosponsors.† This Act would establish the National Drought Council and would have required the Council to

- Develop a national drought policy action plan

---

* 109th US Congress: H.R. 1386. National Drought Preparedness Act of 2005. Sponsor: Rep Hastings, Alcee L. [D-FL-23] (introduced 3/17/2005), Cosponsors: 36, Committees: House Agriculture; House Resources; House Transportation and Infrastructure. Latest major action: 3/24/2005, referred to House subcommittee. Status: Referred to the Subcommittee on Department Operations, Oversight, Nutrition and Forestry, http://thomas.loc.gov/cgi-bin/query/D?c109:1:./temp/~c109AEzi0P.

† 109th US Congress: S.802. National Drought Preparedness Act of 2005. Sponsor: Sen. Domenici, Pete V. [R-NM] (introduced 4/14/2005). Cosponsors: 14. Committees: Senate Agriculture, Nutrition, and Forestry. Latest major action: 4/14/2005, Referred to Senate committee. Status: Read twice and referred to the Committee on Agriculture, Nutrition, and Forestry, http://thomas.loc.gov/cgi-bin/query/D?c109:3:./temp/~c109zrwOqM.

- Evaluate federal drought-related programmes
- Coordinate and prioritize enhancement of the national integrated drought system
- Provide for the development of a drought information delivery system, drought planning models, and drought preparedness plans

The National Drought Preparedness Act of 2005 would also have required the Secretary of Agriculture to establish the National Office of Drought Preparedness and would create in the Department of Agriculture the Drought Assistance Fund, among other things. That Act also would have authorised the states, Indian tribes, local governments, and regional water providers to develop and implement drought preparedness plans. The Act never came to a vote in either house of Congress. The following year, the passage of the important National Integrated Drought Information System (NIDIS) Act of 2006 (Public Law 109-503) provided authorisation for the much-needed drought monitoring systems (see Chapter 10). Most of the recommendations of the 1998 National Drought Policy Commission remain dormant.

Of course, Congress is not the only institution that promulgates policy in the United States. At the federal level, the President and executive branch agencies may also establish certain kinds of policy through executive orders or through the agency rule-making process. For example, the recent issuance of Presidential Policy Directive 8 (PPD-8), 'National Preparedness', which addresses the predisaster preparations for catastrophic natural disasters, including drought, is an example of more holistic disaster policy thinking. In addition, the Department of Homeland Security recently released its National Disaster Recovery Framework (US Department of Homeland Security Federal Emergency Management Agency n.d.), which addresses issues of recovery from a wide spectrum of natural disasters, also including drought. Several other US federal departments and agencies sponsor programmes or policies aimed at some aspect of drought in the United States. For example, the US Department of Agriculture has long sponsored programmes to help the agricultural community recover from drought, but these policies are generally in the nature of crisis management (Wilhite 1986). Drought policy may also be promulgated at the state level, and currently 45 states have some form of drought policy or plan (National Drought Mitigation Center n.d.).

The Use of Science in Policy Making

As described previously, policy in the United States may be developed through the legislative action of federal or state legislatures, through executive orders of the president or state governors, or through a rule-making process by federal, state, or local government agencies. Most public policy is intended to be a solution to a question or problem, achieved through a process of debate and compromise that examines a variety of proposed outcomes and consequences. Whether the subject is tax law, transportation, disaster mitigation, or any other topic, rarely do policy makers have the quality or quantity of data available to make a policy decision a 'no brainer'. The final shape of a policy often results from persuasion among decision makers, each of whom has limited information. Therefore, some uncertainty about effectiveness and impact of a policy almost always remains, even after a policy is formalised.

Most US policy-making bodies, at all levels of government, consider some kind of scientific information in the formulation of drought policy, though scientific information is used in different ways and to different degrees in each case. Evidence-based policy (see Chapters 7 and 9) is being used more frequently within the US policy community. What is generally missing in US policy analysis is the systematic collection of data for the express purpose of evaluating policy success. The US policy community is flooded with evaluative white papers from 'think tanks' that purport to analyse the effectiveness of existing policies, but these studies rarely contain data appropriate for an evidence-based analysis. The United Kingdom and certain other countries seem to have advanced evidence-based policy research further than the United States. See, for example, the compendium of principles and examples by the UK National Endowment for Science, Technology and the Arts (NESTA) (2011).

There has been little inclination in the US Congress, however, to apply those analytical methods to federal policy, mostly because of the time and resource requirements for such analyses, but also because the principles of evidence-based policy are not widely recognised in the halls of Congress. Some progress has been made in Congress in measuring accountability of programmes and policies since the Government Performance and Results Act of 1993 (The White House). Attempts to improve evidence-based accountability are revealed in a study conducted by the US Government Accountability Office (GAO) in 2009 that evaluated methodologies proposed by the Coalition for Evidence-Based Policy (Coalition for Evidence-Based Policy n.d.) for examining social science policy (US Government Accountability Office 2009). In fact, GAO, which is an investigative agency of Congress, has been a strong advocate for establishing measurable goals in federal programmes and then establishing an assessment method for evaluating progress toward those goals. But the US Congress continues to rely largely on the historical process of gathering anecdotal evidence and analysis through expert testimony in hearings for evaluating existing programmes, for the development of new policy, and for incremental amendment to existing policy. When the policy topic is scientific or technical in nature, scientists and technical experts are often invited to testify concerning the technical aspects of a problem or proposed policy.

The US executive branch of government has moved toward evidence-based policy under the Obama administration (Brookings Institution n.d.), and has initiated efforts to apply evidence-based programme evaluations to programme funding decisions (The White House n.d.). Under the George W. Bush administration, related efforts were begun to apply evidence-based policy to a broad area of science policy. The 'science of science policy' was defined this way:

> The Science of Science Policy (SoSP) is an emerging interdisciplinary and international field of research and community of practice that seeks to develop theoretical and empirical models of the scientific enterprise. The development of a strong science of science policy can enable policymakers and researchers to use an evidence-based platform to assess the impacts of the Nation's scientific and engineering enterprise, to improve their understanding of its dynamics, and to evaluate potential future outcomes. Examples of research in the science of science policy include models to understand the production of science, qualitative and quantitative methods to estimate the impact of science, and processes for choosing from alternative science portfolios. SoSP's advocates envision that in the near future the Federal government will be able to utilize science of science policy research to make better policy decisions based on empirically validated hypotheses and informed judgment. (National Science and Technology Council n.d.)

That effort, championed most ardently by presidential science advisor John Marburger, resulted in the formation of a subcommittee under the National Science and Technology Council, which produced a roadmap for evidence-based science policy (National Science and Technology Council n.d.). Although focussed largely on the social sciences, this effort was intended to evaluate science policy of all kinds (Fealing et al. 2011). The National Science Foundation, long interested in the objective evaluation of policy and programme funding efforts, maintains an ongoing programme to research evidence-based science and innovation policy (National Science Foundation n.d.). The National Academies have also recognised the need for evidence-based policy analysis as exhibited, for example, in the National Council of Medicine's report on the efficacy of treatment methods for mental disorders (O'Connell, Boat, and Warner 2009). Evidence-based policy efforts are also pursued by some US states. For example, the Washington State Institute of Public Policy has developed a four-step process for evidence-based analysis of public policy (National Endowment for Science Technology and the Arts [NESTA] 2011).

The Changing View of Science by US Policy Makers

Beyond specific scientific issues or methodologies, some broader fundamental differences may shape the way scientists and politicians approach an issue. Scientists collect data, formulate a working hypothesis, interpret and test the data against the hypothesis, and draw conclusions that might change if new data arrive. Because the scientific method is one of incremental advance, science always contains some element of uncertainty or incompleteness. Ideas and hypotheses are continually tested and examined, and sometimes changed. It is sometimes difficult to communicate the nature of this inherent uncertainty to policy makers. Evolving interpretations under changing conditions looks like waffling or 'flip-flopping' to politicians. Furthermore, scientists often discuss outcomes in probabilistic terms, rather than deterministic terms. Expression of a possible outcome as a probability violates the policy makers' need for conclusive facts on which to base policy. Besides, some argue, the weather forecasters communicate in probabilities and they are often wrong.

Most politicians in the United States are trained as lawyers (with numerous very notable exceptions, of course) and are trained to gather evidence to support their position. Politicians often take a position and look for evidence to support it, like a lawyer defending a client. Political positions are often adopted from constituents or political leaders and then defended against contradictory data or positions, even though gathering evidence after taking a position may seem backward to a scientist. Politicians know that legislative language becomes law and must be enforceable; every law cannot be a flow diagram with scenarios and multiple contingencies. Of course, not all politicians are lawyers, and politicians may change their position in the face of strong evidence, but this fundamental difference in worldviews between scientists and politicians provides many opportunities for misunderstanding between the two groups. But strong consideration of science in policy is important, and both scientists and policy makers must work to bridge the gap between the two mind-sets.

Until recently, science and scientists enjoyed the confidence of decision makers in the United States. Vannevar Bush, the de facto head of US government science during much of World War II, proposed that science and engineering were critical to economic growth (Bush 1945), and the growth of the US economy until recently was attributed substantially to advances in science, engineering, and medicine. However, over the last few years, a growing number of policy makers, most notably from the conservative right wing of US politics, have expressed scepticism, not only on the results of science but also on the process of science itself and how science is conducted in a landscape where key scientific discoveries can cause huge shifts in culture and the economy. Numerous books, articles, and editorials have sought to explain the decline in confidence in science among certain members of policy communities over the last two decades. Though the causes are debated, the rise of scepticism toward science by both policy makers and the public, as further discussed in Chapters 2 and 3, is widely observed and is affecting the role of science in policy making.

The collision of objective data and their interpretation with values and ideology have often produced a combustible mixture, and we are now seeing new examples. For example, the science behind human genomics has collided with the values of individuals who would like to limit the medicinal use of the techniques and findings. Even if a scientist objectively informs policy makers of the pros and cons of a policy decision based on scientific interpretation and experience, nothing requires the politician to accept the information provided. It is becoming more difficult to build political consensus in the United States based on scientific input because of the increasingly vocal and strident scepticism expressed toward science, and because a certain number of people completely reject the validity of scientific observation and interpretation. These sceptics do not accept the methodology and conclusions of science any more than an atheist accepts the authority of Scripture.

The Changing Climate of US Drought Policy

One area of science that has come under attack more frequently and more fiercely than others recently is the science of climate change, as discussed in Chapter 3. Scepticism among some politicians toward the science of climate change has profoundly affected any policy field that is affected by climate change. This includes drought policy, of course. Any attempt to move drought policy beyond the crisis-centred reactive policy of crop insurance requires the acknowledgement that drought is a part of the natural hydrologic cycle and that conditions of climate change may alter the geography, frequency, and severity of drought in the future. As stated in the Synthesis Report of the last Intergovernmental Panel on Climate Change, 'Drought-affected areas are projected to increase in extent, with the potential for adverse impacts on multiple sectors, e.g., agriculture, water supply, energy production and health' (Pachauri and Reisinger 2007).

Formulation of effective drought policy would be difficult even in the absence of climate change, but the shape of effective drought policy will be strongly affected by the uncertain future under changing climate. For example, a very powerful aspect of climate change that should affect any policy related to weather or climate is the end of stationarity. Stationarity is 'the idea that natural systems fluctuate within an unchanging envelope of variability' (Milly et al. 2008, 573). In other words, though the properties of temperature, precipitation, river flow, and other characteristics of natural systems might vary over periods of days, months, or seasons, the long-term conditions remain basically the same. Under conditions of climate change, we can no longer assume stationarity; conditions for a particular geographic area may change monotonically over time, leading to very different conditions than those to which the local population is accustomed. Policy based on an assumption of climatic stationarity may produce very unsatisfactory results in the future.

One example of policy unfortunately based on the principle of stationarity was the Disaster Financing and the Budget Control Act of 2011. The president signed the Budget Control Act of 2011 (Public Law 112-25) on August 2, 2011. The purpose of this legislation was primarily to increase the US debt limit, establish caps on the annual appropriations process over the next 10 years, and to create a Joint Select Committee on Deficit Reduction instructed to develop a bill to reduce the federal deficit over the 10-year period. One provision of this law that is relevant to US policy under climate change and the lack of stationarity was a provision that determined that disaster relief appropriations each fiscal year would be based on 'the average funding provided for disaster relief over the previous 10 years, excluding the highest and lowest years'. In that bill, 'the term "disaster relief" means activities carried out pursuant to a determination under section 102(2) of the Robert T. Stafford Disaster Relief and Emergency Assistance Act (42 USC. 5122(2))', which would include drought.

Using the 10-year average disaster funding formula, which excludes the highest and lowest year, calculation for disaster funding available for fiscal year 2012 (for example) would exclude the emergency funding paid out in 2005, the year that Hurricanes Katrina and Rita hit the US Gulf Coast. Using historical cost data will exclude the likely increase in annual damages indicated by current trend data (Kunreuther 1996) which show disaster losses increasing for a variety of social and economic reasons, and by any increase in the number and/or severity of weather-related disasters resulting from climate change (Karl and Melillo 2009). The overall result of the Budget Control Act will be to chronically underfund federal disaster recovery funding in the United States. The impact of this policy over time will be additional economic burden on state and local budgets and on individuals, or the need for special supplemental appropriations, which will be quite difficult under the guidelines of the Budget Control Act. Though the impact will be felt in the wake of all major disasters, this law constitutes an example of the kind of policy that does not reflect scientific findings or recommendations.

Broader Aspects of US Disaster and Drought Policy

Another aspect of drought policy related to the role of science and technology in mitigating the impacts of drought is the way that the United States addresses disaster policy more generally. US disaster policy has long focussed on the response and recovery phases, rather than the prevention or mitigation of disasters through the strengthening of communities and the improvement of national resilience in general. The Stafford Act has been the centrepiece of US disaster policy since it originally passed into law in 1988.[*] The most widely known disaster law and the most widely cited in the context of traumatic incident, the Stafford Act is intended to 'provide an orderly and continuing means of assistance by the Federal Government to State and local governments in carrying out their responsibilities to alleviate the suffering and damage which result from such disasters...'

Clearly, the Stafford Act is primarily a guide for providing resources in response to disaster incidents. In fact, the Stafford Act was never intended to be a mechanism for preventing disasters. The word 'prevention' is used only once in the Stafford Act. The Stafford Act focuses strongly on providing assistance to help recover from a disaster rather than preventing disasters. Recovery resources are certainly needed because even mitigation measures cannot prevent all disasters, but the value proposition of disaster mitigation or prevention is not well developed at this time and should receive more attention as a way to avoid disasters and reduce the loss of money and human life.

As mentioned before, the current administration has taken some significant steps to address the mitigation or prevention of disasters that might contribute to the long-term resilience of the nation. On March 30, 2011, President Barack Obama signed a new presidential policy directive (PPD-8) with the title 'National Preparedness', which begins by saying: 'This directive is aimed at strengthening the security and resilience of the United States through systematic preparation for the threats that pose the greatest risk to the security of the Nation, including acts of terrorism, cyberattacks, pandemics, and catastrophic natural disasters' (The White House and the Department of Homeland Security 2011).

The directive replaces Homeland Security Presidential Directive 8 (HSPD-8), which was delivered in 2003 and HSPD-8 Annex I, which came out in 2007. The new directive calls for development of a National Preparedness System to guide activities that will enable the nation to achieve the goal, a comprehensive campaign to build and sustain national preparedness, and an annual National Preparedness Report to measure progress in meeting the goal. As stated in the directive: 'The Secretary of Homeland Security shall coordinate a comprehensive campaign to build and sustain national preparedness, including public outreach and community-based and private-sector programmes to enhance national resilience, the provision of Federal financial assistance, preparedness efforts by the Federal Government, and national research and development efforts'.

---

[*] The Robert T. Stafford Disaster Relief and Emergency Assistance Act (Public Law 100-707), signed into law on November 23, 1988, amended the Disaster Relief Act of 1974 (Public Law 93-288). The Stafford Act constitutes the statutory authority for most federal disaster response activities especially as they pertain to the Federal Emergency Management Agency (FEMA) and FEMA programmes.

This directive recognises that US national responses to a wide range of events, from the 2009-H1N1 pandemic to the 2010 BP Deepwater Horizon oil spill to devastating drought, have been strengthened by leveraging the expertise and resources that exist in the community. The Department of Homeland Security is directed to coordinate a 'comprehensive campaign' to reach the goals of the directive. That campaign should be informed by long-term research and development efforts. Science and technology are expressly included as a means to address natural disasters through risk assessment, forecasting, mitigation, and prevention.

The issuance of PPD-8 is a significant advance in moving the federal role in disasters beyond response funding, but the movement initiated by the directive was further enhanced by the report of the Homeland Security Advisory Council's Community Resilience Task Force (CRTF) (Homeland Security Advisory Council 2011). The report, released in June 2011, builds on the Quadrennial Homeland Security Review Report (2010) and contains a set of recommendations intended to define the role of the Department of Homeland Security in advancing national resilience through the mechanism of PPD-8:

> The Department of Homeland Security (DHS) clearly has an important role to play in building national resilience, but at its core, the resilience charge is about enabling and mobilizing American communities. The CRTF acknowledges that many relevant activities are already underway, particularly in fostering development of preparedness capabilities, but observes that those activities are rarely linked explicitly to resilience. Thus, the Task Force identified an urgent need for clear articulation of the relationships and dependencies between resilience and other homeland security efforts—particularly preparedness and risk reduction. Clarification of these relationships is crucial both to build shared understanding across diverse stakeholder communities and to motivate action throughout the Nation. A more integrated approach also offers opportunities to identify and leverage synergies across programs, enabling conservation of scarce resources. (Homeland Security Advisory Council 2011)

The recommendations that call for clarification of responsibilities, building knowledge and public awareness, and providing long-term federal assistance to local communities are especially valuable. Whether—or to what extent—these recommendations will be followed in a climate of competing priorities and diminishing resources remains to be seen. These recent developments are encouraging for the science community, however, because there is recognition that risk assessment, forecasting, mitigation, and preparedness can be enhanced by improving our understanding of disasters like drought, and by exercising the value proposition that it is less expensive to communities and to the nation to reduce the impact of drought disasters through adaptation and prevention than simply standing by while repeated droughts devastate crops and forests and then send out relief cheques.

Changes are also occurring within US agencies with drought relief programmes. The agricultural community has been aided during periods of drought by federal relief programmes such as those administered by the US Department of Agriculture's Farm Service Agency[*] and the federal Crop Insurance Program,[†] which have long provided financial support for farmers caught in the path of a drought or other disasters. The US Department of Agriculture is examining alternative ways of addressing the impacts of drought on agriculture, as demonstrated in a recent report commissioned by USDA that examines the impacts of climate change on the Crop Insurance Program (Beach and Zhen 2010). The United States may not yet be ready for commercial ecological farming (Tirado and Cotter 2010), but much greater progress will be required in adapting agricultural methods to avoid the disastrous effects of more frequent droughts. Policies and programmes enlightened by scientific understanding of drought and climate change will provide more effective defence against drought in the future.

Conclusions

Ultimately, the United States needs a realistic view of future drought probabilities based on a rigorous understanding of climate trends so that individuals, communities, and government institutions can adjust their expectations and strategies to develop meaningful drought strategies and policies. Drought policy based on science will improve risk assessment, monitoring, forecasting, warning, mitigation, and preparedness, and it will allow communities, especially those that rely most strongly on agriculture and forestry, to take adaptive measures to avoid the most severe consequences of drought. In some cases, local, regional, and national leaders may conclude that broad changes are needed in agricultural practices for a particular locale, or that the old practices are no longer viable in a changing world. Providing assistance to help communities adapt to climate change would be far more productive than continuing to send out disaster relief cheques for drought losses. An excellent beginning for this process would be to dust off the recommendations of the 1998 National Drought Policy Commission, which are still valid today and resonate with the principles of modern policy in a world with a changing climate.

---

[*] Farm Service Agency (originally called the Farm Security Administration), US Department of Agriculture, http://www.fsa.usda.gov/FSA/webapp?area=about&subject=landing&topic=ham-ah

[†] Risk Management Agency, US Department of Agriculture, http://www.rma.usda.gov/aboutrma/what/history.html

References

Beach, R. H., and C. Zhen. 2010. *Climate change impacts on crop insurance*. Prepared for USDA Risk Management Agency, by RTI International, Research Triangle Park, NC. http://www.usda.gov/oce/climate_change/files/ImpactsCropInsurance062010.pdf

Brookings Institution. n.d. Available from http://www.brookings.edu/articles/2011/04_obama_social_policy_haskins.aspx

Bush, V. 1945. *Science—The endless frontier: A report to the president by the director of the Office of Scientific Research and Development*. Washington, DC: United States Government Printing Office. http://www.nsf.gov/about/history/vbush1945.htm

Coalition for Evidence-Based Policy. n.d. Available from http://coalition4evidence.org/wordpress/

Department of Homeland Security. 2010. *The quadrennial Homeland Security Review Report: A strategic framework for a secure homeland*. http://www.dhs.gov/xlibrary/assets/qhsr_report.pdf

Fealing, K. H., J. I. Land, J. H. Marburger, III, and S. S. Shipp, eds. 2011. *The science of science policy: A handbook*. Stanford, CA: Stanford University Press.

Folger, P., B. A. Cody, and N. T. Carter. 2010. *Drought in the United States: Causes and issues for Congress*. Congressional Research Service Report RL34580.

Homeland Security Advisory Council. 2011. *Community Resilience Task Force report*. Department of Homeland Security. http://www.dhs.gov/xlibrary/assets/hsac-community-resilience-task-force-recommendations-072011.pdf

Karl, T. R., and J. M. Melillo, eds. 2009. *Global climate change impacts in the United States*. Cambridge: Cambridge University Press.

Kunreuther, H. 1996. Mitigating disaster losses through insurance. *Journal of Risk and Uncertainty* 12:171–187.

Milly, P. C. D., J. Betancourt, M. Falkenmark, R. M. Hirsch, Z. W. Kundzewicz, D. P. Lettenmaier, and R. J. Stouffer. 2008. Stationarity is dead: Whither water management? *Science* 319 (5863):573–574.

National Drought Mitigation Center. n.d. Available from http://drought.unl.edu/Planning/PlanningInfobyState/DroughtandManagementPlans.aspx

National Drought Policy Commission. 2000. *Preparing for drought in the 21st century*. Available from http://govinfo.library.unt.edu/drought

National Endowment for Science Technology and the Arts (NESTA). 2011. *Evidence for social policy and practice: Perspectives on how research and evidence can influence decision making in public services*. UK. http://scienceofsciencepolicy.net/system/files/attachments/Expert_Essays_webv1.pdf

National Science Foundation. n.d. Available from http://www.nsf.gov/funding/pgm_summ.jsp?pims_id = 501084

National Science and Technology Council. n.d. Available from http://scienceofsciencepolicy.net/?q = node/5.

O'Connell, M. E., T. Boat, and K. E. Warner. 2009. *Preventing mental, emotional, and behavioral disorders among young people: Progress and possibilities*. Committee on Prevention of Mental Disorders and Substance Abuse among Children, Youth and Young Adults. Washington, DC: National Research Council and Institute of Medicine.

Pachauri, R. K., and A. Reisinger. 2007. *Climate change 2007: Synthesis report, contribution of working groups I, II and III to the Fourth Assessment Report of the Intergovernmental Panel on Climate Change.* Geneva Switzerland: United Nations Intergovernmental Panel on Climate Change. http://www.ipcc.ch/pdf/assessment-report/ar4/syr/ar4_syr.pdf

Tirado, R., and J. Cotter. 2010. *Ecological farming: Drought-resistant agriculture.* The Netherlands: Greenpeace International. greenpeace.org

US Department of Homeland Security Federal Emergency Management Agency. n.d. Available from http://www.fema.gov/recoveryframework/

US Government Accountability Office. 2009. *Program evaluation: A variety of rigorous methods can help identify effective interventions.* http://www.gao.gov/new.items/d1030.pdf

The White House. n.d. Available from http://www.whitehouse.gov/omb/mgmt-gpra/gplaw2m

The White House and the Department of Homeland Security. 2011. Available from http://www.dhs.gov/xlibrary/assets/presidential-policy-directive-8-national-preparedness.pdf

Wilhite, D. A. 1986. Drought policy in the U.S. and Australia: A comparative analysis. *Journal of the American Water Resources Association* 22:425–438.

6

Institutionalising the Science– Policy Interface in Australia

John Kerin and Linda Courtenay Botterill

CONTENTS

Notable differences between the drought science–policy interfaces in Australia and the United States are the far more developed state of drought science and the engagement of scientists in policy development in the United States. As described elsewhere in this volume, the US government has invested in the development of the National Integrated Drought Information System (NIDIS) and related programmes and research and, in particular, the United States now leads the world in drought monitoring. Furthermore, in the United States, scientists have had a greater role as policy entrepreneurs (Kingdon 1995) in promoting a transition from a crisis response to drought to a risk management approach (Botterill forthcoming), which favours 'preparedness over insurance, insurance over relief, and incentives over regulation' (National Drought Policy Commission 2000, v). There has not, however, been matching political will or capacity to take this risk management-based approach through to national policy.

In contrast, in Australia the scientific effort has tended to follow the policy, with the scientists more confined to implementation. The introduction of the National Drought Policy in 1992 was an act of political will by the Commonwealth and state governments in response to political and policy imperatives. Scientific advice has then been sought to provide the necessary information to give effect to policy decisions. Illustrating this tendency, the major national response to agricultural droughts—the exceptional circumstances (EC) programme—was introduced in the absence of a definition of what constituted such circumstances and members of the body appointed by government to advise on the existence of EC were not briefed on this element of their responsibilities at the time of appointment.

This chapter will discuss the role of science in the exceptional circumstances programme and the limited and largely reactive contribution of the government scientists within the Bureau of Rural Sciences (BRS), an institution created to try to bring more science into the policy process. There are reviews of the relevant institutional and policy developments, complemented by the reflections of some key policy actors, including one of the authors of this chapter. The chapter begins by describing the policy framework within which the exceptional circumstances programme emerged, followed by an overview of the development of agricultural research in Australia and the development of the BRS. In some ways the BRS can be seen as an attempt to bring more of the natural sciences into a policy arena where agrarianism (see Chapters 2 and 3), and later neoclassical economics were most influential.

Australia has a long history of independent government economic advisory bodies; one of the oldest being the Australian Bureau of Agricultural and Resource Economics (ABARE), which was created (as the then Bureau of Agricultural Economics) in 1945. In the mid-1980s the Minister for Primary Industries and Energy, one of the authors of this chapter, identified a gap in the capacity of the Australian government in the area of technical scientific expertise and sought to establish a body that was complementary to ABARE, in order to provide more diverse expert advice to policy makers. The BRS was established in 1985–1986, undergoing a name change to the Bureau of Rural Resources in 1987–1988 when minerals and energy became part of an enlarged Department of Primary Industries and Energy and then reverting to the BRS again in the 1990s, and that is the designation used throughout the chapter. The initial focus of the BRS was on rural production sciences and particular management issues, with increasing attention to natural resources management and then drought research.

Australia's National Drought Policy

Australia has had a National Drought Policy since 1992, when the Commonwealth and state governments reached agreement on a new policy approach that redefined drought as a risk to be managed instead of a natural disaster to be endured or ameliorated where possible. This was part of a broader shift in Australian political economy, whereby governments were seeking to reduce the degree to which they directly supported private

businesses, whether those were car companies or farms. Federal Labor governments (1983–1996) drove these changes but had to work with the states
where there were varying degrees of enthusiasm about these market reforms.
Australia is a centralising federation even though the Constitution allocated
specific and limited powers to the Commonwealth with all residual powers
to the states. The Commonwealth has, however, become increasingly dominant with three developments of note:

- The High Court effectively ruled that the Commonwealth has sole
  rights to income tax, which is the real growth tax, so the national
  government receives most of the public revenue but the states have
  most of the direct spending responsibilities, thus making them subject to financial coercion.

- Commonwealth governments have the power to make 'international treaties', which now include environmental treaties, and then
  enforce obligations in regard to resources management on the states.

- As described in Chapter 3, state disagreements over water allocations and management have enabled the Commonwealth government to become the coordinator of resources management of the
  Murray–Darling Basin.

Under Australia's constitution, agricultural production, natural resources,
land and water management, and disaster relief are state government responsibilities. The national government first became involved in the delivery of
disaster relief in 1939 when it provided £1,000* to the Tasmanian government
to assist with relief following severe bushfires in that state. From the 1930s
until the 1960s, Commonwealth involvement in disaster relief was ad hoc but
a policy approach gradually emerged that saw the Commonwealth matching
state spending on personal hardship relief and the restoration of public assets
in the aftermath of disasters. In 1971 this case-by-case cost sharing arrange-
ment was formalised into the natural disaster relief arrangements (NDRA),
which provide a formula for the division of expenses associated with disaster recovery. The NDRA are triggered when a state government declares that
a natural disaster has occurred and, once state spending reaches an agreed
threshold, increased levels of Commonwealth spending are triggered.

Until 1989, drought was covered by the natural disaster relief arrangements; however, in that year the Commonwealth Minister for Finance
announced that drought was to be removed from the NDRA for three
reasons. First, drought was dominating NDRA spending, particularly in
the state of Queensland, and the Commonwealth government was interested in reining this in. Second, there was evidence that the Queensland

---

* Australia decimalised its currency in 1966, moving away from the Australian pound to the
Australian dollar.

government was misusing the declaration process for party political purposes. Third and of importance to this book was that scientific understanding of the determinants of Australian climatic variability, such as the El Niño-Southern Oscillation, was improving and policy makers considered that it was increasingly inappropriate to consider drought as a disaster similar to the other events covered by NDRA, such as cyclones, earthquakes, and bushfires.

Following the decision to remove drought from the NDRA, a Drought Policy Review Task Force was established to

1. Identify policy options that encourage primary producers and other segments of rural Australia to adopt self-reliant approaches to the management of drought
2. Consider the integration of drought policy with other relevant policy issues
3. Advise on priorities for Commonwealth government action in minimising the effects of drought in the rural sector (Drought Policy Review Task Force 1990, vol. 1, 2)

The Task Force reported in May 1990 and its key recommendations formed the basis for Commonwealth–state negotiations over a new drought policy framework. The Task Force recommended that governments implement a National Drought Policy 'as a matter of urgency' (1990, vol. 1, 21). The panel also advised that government assistance in the event of drought

• Be provided in an adjustment context
• Be based on a loans-only policy
• Permit the income support needs of rural households to be addressed in more extreme situations (1990, vol. 1, 18)

In other words, the intention was to consider the impact of drought in the context of structural changes in agriculture, to reduce reliance on 'handouts', and to consider the welfare aspects of drought, as opposed to just the impacts on businesses.

In July 1992, Commonwealth and state ministers with portfolio responsibility for agriculture announced a National Drought Policy (NDP) based on principles of risk management and self-reliance. The policy was implemented through a package of farm programmes that had been reviewed in 1992 and which were focussed on improving farm management and facilitating structural adjustment in the farm sector. These included interest rate subsidies on commercial loans, income smoothing tax measures (farm management bonds and income equalisation deposits—later combined as farm management deposits) and a welfare payment, added as the drought worsened in 1994 (for a detailed description of the background and details of the NDP, see Botterill 2003). An important component of the policy was the concept of 'exceptional circumstances' under which enhanced government support became available to farmers in areas deemed to be experiencing conditions so severe that even the best farm manager could not be expected to cope. These EC provisions quickly came to dominate the policy as the NDP came into effect in 1993 at the beginning of what was to become regarded as the worst drought of the twentieth century.

The NDP has evolved over its nearly two decades of operation. A major three-part review of the National Drought Policy was undertaken in 2008 (Hennessy et al. 2008; Kenny et al. 2008; Australia. Productivity Commission 2009). At the time of writing, the Commonwealth and state governments were considering their response to the review recommendations. Key changes that have occurred in the implementation of the policy over the past two decades have seen the risk management focus diminished and greater emphasis given to the welfare component of the policy.

The dominance of the exceptional circumstances provisions highlighted the lack of a robust definition of an exceptional circumstance. When the policy was announced, ministers referred to 'severe downturn' and the need to have measures 'permanently in place to avoid implementation of ad hoc policies in times of crisis' (ACANZ 1992, 14). In introducing the enabling legislation into Parliament, the Commonwealth Minister's speech referred to 'exceptional circumstances such as severe drought or substantial commodity price downturns' and 'severe downturns' (Crean 1992) but did not provide any indication as to how such circumstances would be identified. A Rural Adjustment Scheme Advisory Council was set up by legislation, a 'significant role' of which was to advise the Minister 'on request, as to whether exceptional circumstances [were] affecting farmers and whether additional support [was] required in those circumstances' (Crean 1992).

Scientific Input into the Exceptional Circumstances Programme

The paradigm shift in Australia's approach to drought from crisis response to risk management occurred largely at the policy level but advances in scientific understanding of Australia's climate also enabled some rethinking of the policy approach, though the new policy was very much framed by policy makers. As noted before, although based on the principle that drought is a normal part of the farmer's operating environment to be managed like any other risk facing the farm business, the EC provisions were designed to be used when drought conditions became so severe that even the best farm managers could not be expected to cope.* However, ECs were not defined in the legislation or accompanying material in any level of detail. The legislation that set up the scheme contains only two mentions of EC. The first is in the context of the establishment of a Rural Adjustment Scheme Advisory Council (RASAC), one of the functions of which was to 'provide advice to the Minister, after consultation with all interested parties, on matters that the Minister requests advice on and, in particular, whether, under the Rural Adjustment Scheme, exceptional circumstances exist in relation to the farm sector' (Rural Adjustment Act 1992, s8(d)). The second (s21(3)) relates to the division of funding responsibility between the Commonwealth and state governments, should such circumstances be found to exist.

In his speech introducing the relevant legislation into the Parliament, the then-minister did not provide any further information about what would constitute an exceptional circumstance. Throughout the first year of the scheme's operation, 1993, recommendations about the existence of EC were made by the RASAC on a largely ad hoc basis. The first chairman of the Rural Adjustment Scheme Advisory Council, Neil Inall, reports that no council members were briefed on the exceptional circumstances concept prior to their appointments (Interview, October 2011) and yet the first few years of the council's operation were dominated by the assessment of EC claims. The council visited areas that had applied for EC support but did not have a comprehensive, science-based definition or set of triggers to form the basis of decisions.

The work of the RASAC was supported by a Secretariat within the Department of Primary Industries and Energy (DPIE), which in turn drew on the scientific expertise from the BRS. From mid-1994, a principal research scientist from within the BRS, Dr David White, was seconded to the Policy Division of the Department of Primary Industries and Energy to advise policy makers on EC applications. Dr White travelled with RASAC when they visited areas seeking an EC declaration and drew on the resources of agencies such as the Bureau of Meteorology to provide advice on climatic conditions. This advice was somewhat limited as it relied almost entirely on a crude measure of drought–rainfall patterns. Throughout 1995, White sought to have the definition expanded to take account of the impact of temperature on growing conditions. This was achieved with a cabinet decision in September 1995 when a number of decisions to refuse requests for declarations were overturned when attention was given to temperature (Interview, October 2011).

By early 1994, disputes were arising between the Commonwealth government and some state governments about the broad rule of thumb that

---

* It should be noted that the EC provisions were originally designed to apply to any event that fell into this category, not just drought. The first EC declaration was in fact for damage caused by excess rain and there were also EC declarations in the early years of the scheme for the wool industry following a severe slump in the price for wool.

had been applied, which determined that an exceptional circumstances drought was one in which an area had been drought declared 2 years out of 3, with those states wanting to lower that threshold. In October 1994, a workshop was held to review recent studies concerned with providing scientific advice on EC and develop some criteria for the declaration process (White and Bordas 1997). These criteria were considered and endorsed at a meeting of the Ministerial Council later that month. The criteria were very broad, consisting of six core criteria that would be taken into account by Commonwealth and state/territory governments in considering exceptional circumstances declarations:

1. Meteorological conditions
2. Agronomic and stock conditions
3. Water supplies
4. Environmental impacts
5. Farm income levels
6. Scale of the event (ARMCANZ 1994, 3)

In effect, these criteria were starting to consider the broader range of drought impacts as set out in Chapter 1. The framework specified that a rare and severe drought was a 'once-in-a-generation' (ARMCANZ 1994, 8) circumstance, taken to mean a 1 in 20- to 25-year occurrence (O'Meagher, Stafford Smith, and White 2000, 121), with the meteorological situation as the threshold event. This broad definition was amended over time, with ministers revising the criteria in 1999 to three core indicators:

1. The event must be rare and severe.
2. The effects of the event must result in a severe downturn in farm income over a prolonged period.
3. The event must not be predictable or part of a process of structural adjustment.

The 1999 decision also changed the threshold criterion from meteorological conditions, specifying that 'the key indicator is a severe income downturn, which should be tied to a specific rare and severe event, beyond normal risk management strategies employed by responsible farmers...The severe downturn should be for a prolonged period and of a significant scale'. 'Rare' was taken to be an event that occurs 'on average once in every 20 to 25 years but could also include a 'combination of exceptional factors formulating an event' (ARMCANZ 1999, 63).

The process of declaring EC comprised an application from the affected area, forwarded via state governments to the Commonwealth Minister, who referred the application to the successor to the RASAC, the National

Rural Advisory Council (NRAC). Like RASAC, NRAC drew on advice and expertise from within the BRS in arriving at its decision as to whether exceptional circumstances existed in the applicant area. In 2006 the BRS set up the National Agricultural Monitoring System (NAMS), which was intended to provide information to government but also to stakeholder groups to assist them in the preparation of EC applications. NAMS was first considered by the Primary Industries Ministerial Council in 2004 (Primary Industries Ministerial Council 2004) with the full web-based system available from July 2006 (Australia. Productivity Commission 2009, 120). NAMS was primarily established with the needs of government decision makers in mind but 'it was envisaged that the majority of users would be the general public' (Leedman et al. 2008, 70). O'Meagher et al. (1998, 247) argue that 'many farmers are overwhelmed by the magnitude of the available information' and, while NAMS was intended to provide information in an accessible form, the survey results in Chapter 12 reinforce the general view of very limited uptake amongst farmers.

The review of drought policy initiated by the Commonwealth government in 2008 appears to have placed NAMS into a form of limbo. Although $A1.2 million was allocated in the Commonwealth budget for the programme in 2007–2008 (Australia. Productivity Commission 2009, 122), the interface has been removed from the Internet. It is understood that the Australian government, through the BRS, is working on a replacement system, but at the time of writing, nothing has been announced as to the nature or purpose of the new scheme. The case of NAMS again emphasises some important differences between Australian and US approaches to drought monitoring. First, the name of the programme underlines the agriculture focus of drought science and policy in Australia. Second, it demonstrates the fact that science follows policy and that drought monitoring has been thought of narrowly in terms of triggering government support rather than providing timely risk management information to individuals. This latter point was illustrated by the Productivity Commission's (2009, 186) observation that 'the National Agricultural Monitoring System (NAMS) is an information system that is used in the exceptional circumstances application and assessment process. If the recommendations of this report are followed NAMS will, in time, not be needed for this purpose'.

A key challenge for the scientists involved in the policy process has been striking the right balance between transparent and clear indicators of drought and indices that encompass the different factors that make a drought event exceptional (White and Walcott 2009, 607). This is a challenge for any form of drought trigger (Steinemann et al. 2005). Leedman et al. (2008, 70) note of the EC programme that 'many stakeholders viewed the analyses undertaken in the EC assessment process as something of a "black box" and as a result did not always accept the rationales or methods used in the process'.

Scientific advice both broadly, in terms of the refinement of drought indicators and triggers for support, and specifically, in the analysis of individual applications for assistance, has been crucial. Inall reports that RASAC became 'very reliant' on BRS advice. However, he also emphasises that the council needed to balance competing forms of evidence and was conscious that, during the period of Inall's chairmanship (1993–1998), BRS did not have the capacity to provide advice on social issues (Interview, October 2011), something further discussed in Chapter 8. A key problem for the programme, and this was beyond the control of the BRS, was the constant redefinition of EC, leaving the strong impression that the programme was not particularly science based and was open to political negotiation. The relatively short life span of NAMS only serves to reinforce the impression that drought declarations occur at the whim of policy makers who pick and choose their definitions and their indicators.

The Role of the Bureau of Rural Sciences

The scientists within the BRS have not benefited from the relative independence of their counterparts at the National Drought Mitigation Center or at the National Integrated Drought Information System in the United States (see Chapters 10 and 11). The Australians have been more reactive and, as direct government employees, their programmes have been more susceptible to changes by policy makers. In order to understand how this situation came about, particularly in light of the strong policy approach toward risk management as the basis of drought policy and the implied need for good science, it is worth considering the history of agricultural policy making in Australia and the role therein of independent advisory agencies.

Agricultural Policy Making in Australia

As noted earlier, agricultural policy is constitutionally a state government responsibility but the Australian government has been involved through various mechanisms of consultation and through the funding of programmes since the 1930s. From there, policy coordinating institutions developed to link formally the states and Commonwealth. In 1934 the Australian Agriculture Council was established and four issues of importance were discussed:

1. Means to make it possible for Australia to speak with one voice on agricultural and marketing matters
2. Determination of a definite policy in regard to international and especially intra-Empire* marketing relations

---

* This refers to the British Empire, with which Australia still had considerable economic interactions despite its political independence.

3. Formulation of a definite policy on wheat, both immediate and ultimate

4. Finalising of a basis of a rural rehabilitation scheme through relief of farmers' debts (cited in Grogan 1968, 298)

The council was an important forum for debating policy matters of national importance. For example, in the 1950s and 1960s the council discussed industry-specific matters, broader policy direction such as the levels of government intervention in agriculture and the details of export agreements. With the increasing economic pressures on farmers becoming more evident in the late 1960s and early 1970s, the council began also to consider structural adjustment policies and programmes. In the 1980s and 1990s the council oversaw the dismantling of Australia's complex 'bewildering array of policy instruments which directly or indirectly affect[ed] farm prices' (Throsby 1972, 13). The composition and name of the council varied over time as it added New Zealand to its membership in 1992 and expanded its reach to include resource issues in 1993.[*] The council sought additional scientific advice and here Australia's coordinate federalism links with another institutional tendency in the country: the development and use of advisory agencies, such as the BRS.

The Role of Science at the Production Level

Although the drought science and scientists appeared to follow the policy with respect to drought policy implementation, this is against a backdrop where rural scientific research has played a key role in the development of Australian agriculture, from its beginnings as an offshoot of the British Empire. Agricultural research was mainly carried out by state governments until Federation (1901), through a network of research 'stations' (field sites) in farming areas that concentrated on extension and research into areas readily adoptable by producers.

It was not until 1921, when the Institute of Science and Industry was established, that the Australian government became actively involved in agricultural research. The institute later became the Council for Scientific and Industrial Research and, in 1949, the Commonwealth Scientific Industrial Research Organisation (CSIRO). Until the 1960s most research from the institution was with respect to agriculture and forests, with research laboratories spread throughout the continent. After World War II, most private research was associated with agricultural and veterinary chemicals.

---

[*] More recently, it has reverted to an agriculture focus and is currently known as the Primary Industries Ministerial Council (for further information on the operation of the council, its evolution and the issues it has addressed, see Botterill 2007).

Until 1955 there was little direct financial contribution by producers or universities to agricultural research. In Australia, there is no equivalent of the US land grant universities and, while some of the major state universities had agricultural faculties or schools, there were no direct extension functions. In 1955 a policy was adopted by the Australian government to have compulsory levies on production with government grants supplementing funds raised entailing specified conditions for the expenditure of such funds on research. Research committees, dominated by farmers, were established to make recommendations on the allocation of funds by the Minister and projects were mainly carried out by the CSIRO, which received the largest proportion; the states and federal government departments; and some of the universities. Very close ties developed between the commodity-based research committees and the major national statutory marketing authorities, which had the authority to acquire and then market produce, largely as a result of key producers holding roles in both. As a result of a series of government decisions, by 1989 the agricultural research organisations become a system of some thirteen agricultural research and development corporations (RDCs) and five research councils.

As well as the CSIRO, state research facilities, the RDCs, the universities, and some private bodies, there are other institutions active in agricultural research such as cooperative research centres (CRCs) and membership of and involvement in international research bodies. The gradual adoption of regional natural resource management (NRM) by the national and state governments, together with programmes such as Landcare, has also seen the development of NRM policies from the ground up with the formation of fifty-six catchment management authorities with implementation plans and a potential capacity to act as conduits for scientific research results and land and water management techniques. As a result of these initiatives Australian agriculture is well endowed with research capacity and has strong international research linkages. The research and development corporations have influenced policy indirectly as farm organisations, and policy-aware officials have influenced the direction of research. Some of the CRCs have sought to inform policy and, by having departmental officials on the boards, have been able both to inform and influence some policy.

In 1945 the Australian government established the Bureau of Agricultural Economics (BAE), later to become the Australian Bureau of Agriculture and Resource Economics (ABARE). These organisations were institutions of applied agricultural and resource microeconomics, though they have provided a wider range of economic advice to ministers and the government over the years. The researchers are independent in that they can set some of the research agenda and publish as they wish.

The BAE started with a strong focus on production but over time took a greater role in considering efficiency, which was to have considerable policy implications. The first major task of the bureau was to advise on the earning potential of thousands of proposals to settle returning soldiers onto farms so as to avoid the problems of World War I soldier settlement schemes. The second major task was to analyse domestic pricing arrangements for agricultural commodities, following wartime price controls. Farm surveys were initiated and indexes established to support the implementation of domestic pricing arrangements. The Korean War wool boom saw buoyant exports and the Australian government sought ways to encourage the expansion of production and exports, which led the BAE to analyse the economics of profitable production, including issues such as the economics of land clearing, dams, irrigation schemes, and beef roads. The BAE also analysed the implications of the Common Agricultural Policy of Europe in 1957 and US farm policies, well before agriculture became an issue in the Uruguay Round of multilateral trade negotiations in the late 1980s. Annual BAE/ABARE Outlook Conferences, initiated in 1971, had a major policy effect by exposing farm organisations to the economics of agricultural policy.

On the election of the Hawke Labor government in 1983, one of the present authors in his role as the new Minister for Primary Industry announced in his first speech the intention to establish a science equivalent to the BAE. The motivation for this decision was the strong perception that the Department of Primary Industry was heavily directed to economics, trade, and the running of key programmes based on a deep knowledge of the commodity situation, export meat inspection, and a range of support programmes, as well as farm adjustment issues. Convinced of the role that the BAE had played in influencing policy and believing that future issues would, additionally, be more about natural resource and more complex policy questions, the view was that, if a sound structure for a scientific clearing house could be established and people of scientific expertise could be recruited, broader and beneficial policy advice would be forthcoming. The BRS was established in 1986 to fill this role. Perhaps tellingly, though, in 2010, the BRS was amalgamated with ABARE to form the Australian Bureau of Agricultural and Resource Economics and Sciences and the relative activity by type of science will reveal much in a future review.

Evidence-Based Policy Making in Australian Drought Policy

This brief history illustrates not only the importance of scientific research in Australian agricultural policy, but also its heavy dependence on government funding and support. It also highlights the strong focus on the production of science for policy. Although the term 'evidence-based' policy making has become associated with recent policy debates, particularly since the election of the Blair government in the UK with its emphasis on finding 'what works' (Cabinet Office 1999), it can be seen that the Australian government has been setting up expert bodies to provide advice, or evidence for policy, for some decades.

The Role of BRS in Drought Policy Implementation

In retrospect, it seems an incredible oversight that policy makers did not seek scientific advice as to what would constitute an exceptional circumstance under the National Drought Policy from the outset. An important point to be made when considering the role of science in the implementation of the exceptional circumstances programme is that EC was originally an add-on to a programme aimed at facilitating structural adjustment in Australian agriculture. The original National Drought Policy was not—and this remains the case—a single piece of legislation comprising a number of programmes. It is an umbrella policy under which sits a number of government programmes and policies, each of which has its own legislation. These legislative underpinnings have changed over time. This is important because it partially explains the failure to engage scientists in developing a comprehensive definition of EC and the lack of a set of agreed triggers when the policy first commenced.

The original National Drought Policy was to be delivered through three key pieces of legislation: the *Rural Adjustment Act 1992*, the *Farm Household Support Act 1992*, and the *Income Equalisation Deposits Act 1992*. These three pieces of legislation were introduced into the Parliament together and were aimed at improving farmer self-reliance and risk management capacity. The clear intent was that there would be consistency between government objectives relating to drought policy, structural adjustment, and sustainability. The provision of drought support through the rural adjustment scheme ensured that support was only available to farmers who had 'good prospects for long term profitability' (Crean 1992, 2414). Even the farm household support programme included strong incentives for farmers who were facing difficulties to leave the land. In combination, the programmes were designed to ensure that drought support did not impede the structural adjustment process by supporting unviable farm businesses. The exceptional circumstances provisions were intended to be used when conditions were just that: exceptional. Neil Inall reports that he accepted the Minister's invitation to chair RASAC because he had a long-standing interest in structural adjustment; he probably would not have joined the council had he known its activities would be so dominated by the determination of eligibility for drought relief (Interview, October 2011).

Against this background, the evidence that was anticipated to be the most important for RASAC to undertake its role was economic data emanating from ABARE on the agricultural industries. The need for extensive and regular drought monitoring, indicators, and scientifically based triggers was

not foreseen. It very quickly became apparent, however, that the EC programme was going to be needed because of the deepening drought from
1993 onwards and it was at this point that BRS became engaged in providing
scientific advice.

Since the early 1970s, Australian agricultural policy has been dominated
by the general trend of the dismantling of government assistance to farming.
This accelerated in the 1980s and the programmes that had been in place,
in the form of stabilisation schemes, home consumption schemes, statutory
marketing arrangements, guaranteed minimum process, export monopo-
lies and a variety of other market interventions, were phased out. These
moves were informed by the research of ABARE, the Industries Assistance
Commission, and its successors and an increasingly 'dry' economic approach
within the Department of Primary Industries and Energy itself. As Minister,
one of the present authors encountered resistance from the department to
the establishment of the BRS, and as an adviser to one of his successors,
the other of us saw the antagonism from officials within the Department of
Primary Industries and Energy toward the introduction of a drought welfare
programme that was not linked to farm viability. In addition, during the
time of the development of the National Drought Policy, the scientists in the
BRS were in many ways the 'poor cousins' to their colleagues in the larger,
economically focussed ABARE.

Where ABARE has over many years developed its own research capacity, employing economists with doctorates and considerable expertise in
their respective fields, the BRS has always had more of a role as coordinator
and collector of the work of others; it has acted as an intermediary between
the scientific research community and the policy process. As former BRS
scientists, Walcott and Clark (2001, 3) note of BRS's role in determining the
existence of exceptional circumstances: 'We saw our role as presenting the
facts, interpreting the likely cause and effects, and warning of possible consequence, within a framework that allowed decisions to be made'. However,
this apparent scientific detachment was clearly difficult for the scientific
advisors to sustain. The authors go on to state (2001, 4):

> At times we found it a struggle to separate giving objective assessments
> from an emotional response to the situation that many rural commu-
> nities were experiencing. Maintaining scientific objectivity is a tenu-
> ous state. Although other players were making the difficult decisions,
> we could not escape the responsibilities that interpreting the advice
> imposed.

This reflects one of the challenges for the expert adviser when his or her
input is sought into an emotionally fraught and values-laden policy debate.
Not only, as Pielke (2007) notes, do scientists have reduced influence where
values are important, but there is clearly also potential for participants to be
affected personally by the situation.

Conclusion

The Australian government has a long history of creating specialist advisory bodies employing disciplinary experts to provide evidence for the policy process. The Australian Bureau of Agricultural and Resource Economics and Sciences of today has provided advice on agricultural economics for over 60 years; the Productivity Commission has its origins in the Industries Assistance Commission established in 1974 and there has been a range of other bodies established, some of which have survived and others that have not. The Bureau of Rural Sciences was established to inject some scientific expertise into the policy deliberations of the Department of Primary Industries and Energy. This review, however, notes some limitations of that organisation. It was a broker of science, rather than undertaking fundamental scientific research, and as such, may not have had the independence and policy entrepreneurship evident in the economic advisory bodies. Second, the science was clearly being sought to inform the development of policies framed around neoclassical economic theory relating to the efficiency of competitive and unfettered markets. Whether this subordination is a function of the eminence of economic science in policy development or the origins of different organisations is an area for further research. Third, the application of science to drought policy may have been seen as a phase while resources managers became more self-reliant. Then the emphasis would presumably be more on meteorology and forecasting.

The role of the BRS with respect to drought policy has only been a small part of its remit. In 1994, there was only the one officer on secondment to the Rural Policy Division to assist with advising RASAC (Interview, October 2011). When he retired in 1996, a small unit was established within BRS to continue providing advice on EC applications. As mentioned previously, in 2010 the BRS was absorbed into the (slightly) renamed ABARES. It is not clear what this will mean for either the scientific work or, more specifically, the drought advice that the bureau has undertaken. Announcements by ministers imply that the revised drought policy currently under consideration will not include EC declarations, potentially diminishing the need for drought monitoring and certainly removing the need to develop triggers for support. It remains to be seen whether the replacement of the National Agricultural Monitoring System moves away from drought advice for government and closer to the US model of drought advice for individual risk managers.

References

ACANZ. 1992. *Record and resolutions: 138th meeting,* Mackay 24 July 1992. Commonwealth of Australia.

ARMCANZ. 1994. *Record and resolutions: Fourth meeting,* Adelaide, 28 October 1994. Canberra: Commonwealth of Australia.

———. 1999. *Record and resolutions: Fifteenth meeting,* Adelaide 5 March 1999. Canberra: Commonwealth of Australia.

Australia. Productivity Commission. 2009. *Government drought support.* Report No 46. Melbourne.

Botterill, L. C. 2003. Uncertain climate: The recent history of drought policy in Australia. *Australian Journal of Politics and History* 49 (1): 61.

———. 2007. Managing intergovernmental relations in Australia: The case of agricultural policy cooperation. *Australian Journal of Public Administration* 66 (2): 186–197.

———. Forthcoming. Are policy entrepreneurs really decisive in achieving policy change? The development of drought policy in the USA and Australia. *Australian Journal of Politics and History.*

Cabinet Office. 1999. *Modernising government white paper.* CM4310. London: TSO.

Crean, S. 1992. *Rural Adjustment Bill 1992: Second reading speech.* Parliamentary debates: House of Representatives Hansard, 3 November 1992.

Drought Policy Review Task Force. 1990. *National drought policy.* Canberra: Commonwealth of Australia.

Grogan, F. O. [1958]. 1968. The Australian Agricultural Council: A successful experiment in Commonwealth–state relations. In *Readings in Australian government,* ed. C. A. Hughes, 297–317. St Lucia: University of Queensland Press.

Hennessy, K., R. Fawcett, D. Kirono, F. Mpelasoka, D. Jones, J. Bathols, P. Whetton, M. Stafford Smith, M. Howden, C. Mitchell, and N. Plummer. 2008. *An assessment of the impact of climate change on the nature and frequency of exceptional climatic events.* Canberra: Bureau of Meteorology and CSIRO.

Kenny, P., S. Knight, M. Peters, D. Stehlik, B. Wakelin, S. West, and L. Young. 2008. *It's about people: Changing perspectives on dryness—A report to government by an expert social panel.* Canberra: Commonwealth of Australia.

Kingdon, J. 1995. *Agendas, alternatives and public policies,* 2nd ed. New York: Longman.

Leedman, A., S. Bruce, and J. Sims. 2008. The Australian National Agricultural Monitoring System—A national climate risk management application. *Proceedings of the USDA/WMO Workshop on Management of Natural and Environmental Resources for Sustainable Agricultural Development,* ed. R. Stefanski and P. Pasteris. Portland, OR: World Meteorological Organization. Technical Bulletin WAOB-2008, NRCS-2008, and AM-10, WMO/TD No 1428. http:// www.wmo.int/pages/prog/wcp/agm/publications/agm10_en.php (13–16 February 2006).

National Drought Policy Commission. 2000. *Preparing for drought in the 21st century.* Available from http://govinfo.library.unt.edu/drought

O'Meagher, B., L. G. du Pisani, and D. H. White. 1998. Evolution of drought policy and related science in Australia and South Africa. *Agricultural Systems* 57 (3): 231–258.

O'Meagher, B., M. Stafford Smith, and D. H. White. 2000. Approaches to integrated drought risk management: Australia's National Drought Policy. In *Drought: A global assessment,* ed. D. A. Wilhite, 115–128. London: Routledge.

Pielke, R. A., Jr. 2007. *The honest broker: Making sense of science in policy and politics.* Cambridge: Cambridge University Press.

Primary Industries Ministerial Council. 2004. *Primary Industries Ministerial Council communique*. PIMC 6, 27 July 2004.

Rural Adjustment Act. 1992.

Steinemann, A. C., M. J. Hayes, and L. F. N. Cavalcantus. 2005. Drought indicators and triggers. In *Drought and water crises: Science, technology, and management issues*, ed. D. A. Wilhite, 71–92. New York: CRC Press.

Throsby, C. D. 1972. Background to agricultural policy. In *Agricultural policy: Selected readings*, ed. C. D. Throsby, 13–22. Ringwood: Penguin Books.

Walcott, J. J., and A. J. Clark. 2001. Risk assessment and expert opinion in implementing policy—Exceptional circumstances. *Proceedings of the 10th Australian Agronomy Conference*, at Hobart.

White, D. H., and V. Bordas, eds. 1997. Providing scientific advice on drought exceptional circumstances. *Proceedings of a Workshop on Indicators of Drought Exceptional Circumstances*. Canberra: Bureau of Rural Sciences.

White, D. H., and J. J. Walcott. 2009. The role of seasonal indices in monitoring and assessing agricultural and other droughts: A review. *Crop and Pasture Science* 60:599–616.

7

Scientific Research in the Drought Policy Process

Roger Stone and Geoff Cockfield

CONTENTS

It has been argued that 'evidence-based policy making should work well' (Choi et al. 2005, and Chapter 9, this volume) because it is based on the assumption that better information will lead to better policy outcomes. In particular, following on from a comprehensive overview by Choi et al., scientists working on drought issues can readily produce various forms of evidence of and about drought, so therefore this should enable policy makers to optimise decision making. An extension of this optimistic view and approach is that policy makers, in turn, provide scientists working on 'drought' with evidence requirements and resources for their research to enhance future policy decisions (Choi et al. 2005). More pessimistically, it has been argued variously that evidence-based policy does not always work, or perhaps almost never works (Parsons 2002; Botterill and Hindmoor 2012), and there are several tendencies that work against it. The natural sciences approach to decision making 'relies excessively on the use of linear systems and discounts non-scientific forms of knowledge' (McLain and Lee 1996). Further to this, scientists engaged in policy development and advice may not be prepared to share their advisory and decision-making roles with other stakeholders or to participate with them on an equal footing.

Scientists may be rather pessimistic about the extent to which their own or other's research is used (see Chapter 12 for more on this), with even the policy makers sceptical about the usefulness of drought research to their needs. The root problem may be that the imperatives that drive scientists and policy makers are quite different, especially in regard to what both groups consider to be 'good evidence' (Choi et al. 2005). Then, there are communications barriers, with scientists speaking their own language consisting of at least some Greek letters and mathematical symbols; this requires translation before it can be understood by nonscientists or even scientists in a different field. Similarly, policy makers also speak their own language riven with acronyms, which are in turn defined by other acronyms. Furthermore, communication amongst policy makers is often for closed audiences and driven by latent political agendas. This defensiveness by policy makers is understandable, given the need to balance the arguments for drought as a natural event against the clamour for action and funding that follows increasingly dry conditions (see Chapters 1, 3, 5, and 9). To add to this complexity, scientists and policy makers work to different time scales where policy makers require answers 'immediately' (Ludwig 2001; Choi et al. 2005) and scientists are inclined to wait for more 'evidence'.

Lofstedt (2005, 25) also suggests that scientists and policy makers often lack trust and respect for the role that each plays. Scientists generally, it is argued, have a lack of respect for those who are not scientists and may resent the power of policy makers to control research funding and the potential for them to misuse scientific data to support a political agenda. This, in turn, may mean that policy makers resent the perceived arrogance, naivety, and, especially, tunnel vision of scientists (Choi et al. 2005). As will be argued in this chapter, scientific research for drought (policy) in Australia seems to conform to and illustrate the 'two-communities thesis' of drought scientists and drought policy makers seemingly unable to take into account the realities and perspectives of one another. One may then wonder how science has a serious role to play in working with farming communities, urban communities, and policy makers in presenting clear but useful scientific information on drought. Key and fundamental scientific roles include properly identifying drought return periods within the context of the many differing scales of climate variability and climate change, juxtaposed against the issue of shorter time horizons in the thinking of many in the policy and general community (for example, Choi et al. 2005), which may then lead to the well-known 'hydro-illogical cycle' in policy responsiveness.

Lowe (2002) argues that many of the issues associated with environmental problems (especially such as drought) are actually the result of applying narrow, specialised knowledge to complex systems. He describes modern science as 'islands of understanding in oceans of ignorance' and calls for 'scientists and practitioners to work together to produce trustworthy knowledge that combines scientific excellence with social relevance'. This also means, it can be argued, that contrary to the intention, if over-reliance on reductionist science in drought research occurs, this can shift the burden of complexity and uncertainty in this type of natural resource policy from the scientists to the policy makers. This statement, together with the arguments presented earlier, forms the key thesis and purpose of this chapter.

The Drought Science/Policy Interface in Australia

Scientists working in the drought arena (and the lead author is one of those), perhaps somewhat arrogantly, believe that 'islands of understanding' exist in the realms of drought-related areas such as climate science, meteorology, agronomy, and hydrology—and perhaps also in the related analyses from economics. Additionally, they also believe 'oceans of ignorance' exist in government bureaucracies, the media and the general public, and the very practitioners who are responsible for provision of emergency relief—especially in those developing appropriate policies in regard to severe drought occurrence (after Lowe 2002). A common sentiment of some meteorologists the lead author is acquainted with is that 'the public and government need to be *educated* to understand the complexities of science since the real problem is ignorance on the part of 'users' of climate and drought science.

To overcome these issues, knowledge brokers (Meyer 2010) or translational scientists would operate as go-betweens for scientists and policy makers. Additionally, it is argued, some organisational capacity that would build interventions would be helpful. In this workplace, developments could occur where scientists would be actively encouraged to enter the policy-making arena. It has been the lead author's experience that the success of such an approach has been mixed, at best. An example from the Australian state of Queensland is illustrative. For over 15 years, two effective climate research institutes existed in Queensland. One was known as the Agricultural Production Systems Research Unit (APSRU), located in Toowoomba at the edge of one of Australia's most prominent agricultural production regions, the Darling Downs. One of the core reasons for the success of APSRU was that it brought together leading systems scientists in agriculture, climate science, hydrology, soil science, and rural extension theory—all drawn from world-class parent bodies such as CSIRO, the Queensland Government's Department of Primary Industries and Natural Resources (as they were then known) plus a key component representing the University of Queensland. APSRU had a strong focus on scientific rigour, strong external publication success, and industry funding for research and development and was responsible for some highly innovative world-class science, which had direct implications for drought research.

Almost in parallel to this, another group existed at a state government research facility in Brisbane, the capital of Queensland, that had a strong focus on providing scientific evidence-based policy, especially in respect to government drought policy delivery, although policy specialists were not based within this group. To add to the complexity of an already complex science-policy environment, a core component of APSRU involving climate science and rural extension was 'hived off' from APSRU by the Queensland Department of Primary Industries in order to provide more direct policy-relevant input into issues such as drought and climate forecast information, including information relevant to long-term issues such as climate change and drought occurrence. This new group coexisted with the Brisbane climate/drought research centre and was known as the Queensland Centre for Climate Applications (QCCA). QCCA enjoyed considerable success, including being awarded the Premier's Award for Excellence in 2005 for its contribution to drought and overall climate understanding by the community and government. Unfortunately, winning such an award is also known in Queensland government circles as a 'kiss of death', presumably because one's research or operational unit is brought so much into the political foreground that it becomes 'ripe for picking' for aspiring government ministers (secretaries) and department heads. Perhaps not surprisingly, QCCA ceased to exist in 2007 as it was swallowed up into a megacentre (known as the Queensland Climate Change Centre of Excellence—QCCCE), which incorporated, rather imaginatively, a major climate/drought policy division from within the Queensland government.

With an initial staffing of over ninety people, this new centre quickly dropped in number to about thirty persons and the policy arm of QCCCE was moved back into a larger environmental policy programme within the Queensland government. A very large component of core scientists resigned shortly after this amalgamation and realignment. Such bureaucratic restructurings are both common and relatively easy to initiate. Agency structures can be changed at the whim of the government executive, which is largely controlled by the lower houses of parliament, both state and federal. With the Australian electoral and party systems, there is usually a clear majority in that lower house and, with relatively strong party discipline, the executive enjoys considerable power. While major departments such as Treasury and Education are reasonably stable, more specialist agencies, such as QCCCE—or, for another example at the federal level, the lead agency on climate change—can be shifted quite easily. Sometimes the restructuring is focussed on the particular agency, shifting to or from a high-profile or central agency; at other times the changes are a product of more general restructuring. The QCCCE case was an example of the former.

It appears that when making such a radical change one also has to consider the variety of personal empires being seen to be destroyed, all the jealousies that arise over 'who is going to be boss' of this new type of integrated centre, and all the differing approaches to science between groups that focus on hard journal publications and external funding needs and a group that had been almost entirely focussed on provision of advice to government, especially on issues such as drought. Also, in such an integrated group that fits nicely the concepts by Choi et al. (2005) of 'organisational capacity building interventions' between science and policy, issues arise as to whether a scientist or a policy specialist should manage such a group and whether or not the policy specialists will be able to mix comfortably with the scientists and really understand the information (and misinformation?) the scientists will deliver to the policy makers in this new, enlightened environment. All of this activity is occurring within a relatively small and potentially easily focussed research and policy centre in a state government regime setting that, on all accounts, should be easily achievable.

Nevertheless, state governments in Australia are subject to policy shifts in regard to aspects such as drought and rural policy, depending on whether there is a change of minister, change of premier or change of political party running that state and hence managing issues related to drought and rural policy. All or any of these changes may mean major changes to the core institutions responsible for provision of science input into drought policy. For instance, a conservative government with significant rural connections may focus on drought issues while a government dominated by political parties with a greater focus on the environment and less support in rural areas may be more interested in strategic climate change policy (see Chapters 4 and 9).

Additionally, in Australia, there has been almost a culture of competition between those 'drought science' and policy agencies in state governments and those in the federal government in Australia. Australia is a centralising federation, even though the Constitution was explicit in the division of powers, giving states responsibility for natural resources, including water and land, as discussed in Chapter 3. The states also control the local governments, at which level there is also land use planning and water management. The increasing power of the federal government can then lead to the states pushing back, especially when the dominant state-level political party is of a different complexion from that in the federal sphere. There is also competition between states. In the science arena, at one stage there was almost open warfare on whose pasture model was the 'best' ('biggest'); therefore, one could not trust the other to provide objective information required for drought severity assessments and exceptional circumstance payments.

This combative environment was exacerbated by the state governments 'going to bat' for all their state rural lobby groups, agencies, and individual land owners they represented in order to obtain the highly prized federal drought relief payments. Due to various allegiances within and between government agencies involved in drought science and policy development, this meant that it was not difficult to create a situation of highly competitive and often counterproductive outcomes where scientific agencies and individual scientists openly mistrusted each other in terms of scientific capability (depending on whether they were supported by either state of federal departments)—let alone contending with the issues associated with the different culture existing between scientists and policy makers in both political environments. No wonder there can easily develop almost a complete lack of cohesive science-based, evidence-based policy when it comes to drought issues in Australia. One wonders how policy agencies were able to develop policy at all in an environment riven with mistrust, envy, and lack of cooperation.

To help overcome many of these issues, the federal government (2006) developed an innovative programme known as the National Agricultural Management System (NAMS) that would bring together all the best scientific outputs from the state governments and the federal government and associated agencies (for example, Bureau of Meteorology, CSIRO, Bureau of Rural Sciences). This now suspended system (see Chapter 6) appears, however, to have gained little recognition amongst farmers (see Chapter 12). This approach appeared to contain all the requirements that would deliver scientific objectivity into the drought policy process. Indeed, a 'science advisory group' was also developed (and proposed by the lead author) that would oversee the processes that would identify the most suitable scientific systems, especially crop, pasture, and climate models that were relevant for Australian (drought) conditions. It is not clear why NAMS did not last. The federal government perspective was that the state governments withdrew their support for such a comprehensive system and retreated back into their own scientific information systems, mainly because they saw no real value in providing financial support to such a system when, to them, it provided no real benefit to a state government. The state perspective is no doubt somewhat different.

Bringing End Users into the Science/Policy Interface

Choi et al. (2005) suggest that scientists and policy makers are not equal partners and therefore additional work needs to be done to improve their combined effectiveness. They make the point that a challenge for scientists if they are interested in influencing policy (which is not always the case and this certainly applies in the experience of the lead author), is not only to convince policy makers of the value of their research and development outputs but also to convince and mobilise the relevant public. In this case this means those people in rural regions prone to drought, usually farmers, who are often on various rural management boards and often have a direct line to important politicians. This could mean influencing public opinion as to the value of various scientific approaches and 'breakthroughs' (for example, crop and pasture simulation models that can reproduce yield values for the past 100 years or more, new developments in seasonal climate forecasting that can help drought preparedness, and new developments in remote sensing) that may have direct benefit on policy settings in relation to drought severity assessment, drought preparedness policy, and so on.

One of the more innovative approaches in Australia in terms of linking scientific research into community decision making and public opinion and, therefore, assisting in drought policy has been the development of climate- and drought-focussed workshops (see Chapter 10 for approaches in the United States). Importantly, where these have succeeded, they have been organised and facilitated by local extension or similar operatives with the climate and drought science purposely provided as an underpinning component to farmer and rural industry decision needs. An example of this is the Managing for Climate and Weather Workshops in Queensland, where many thousands of farmers and those in rural industry paid for and participated in day-long workshops that facilitated understanding of climate science and drought preparedness and also considered opportunities that could be gained in the better years. Through these and other workshops, the lead author has spoken directly as a climate scientist to over 30,000 farmers, mostly in eastern Australia, over the past 20 years. Indeed, a platform of Queensland state government drought policy (Queensland government 1992) was that 50 percent of farmers in Queensland were aware of and including climate forecasting into their decision-making processes. A review by a federal grant funding agency in 2000 established that, indeed, 50 percent of farmers in Queensland were including climate forecasting into their management planning, including drought preparedness planning (CLIMAG 2001; of survey data, Chapter 12). Similar 'success stories' exist in New South Wales, Western Australia, Victoria, and South Australia (Stone and Meinke 2007).

Stone and Meinke (2007) point out that, in regions where there has been successful uptake of more complex climate and weather information by communities, farmers, and industry in drought risk management, these groups participated in the development of appropriate response strategies to climate and weather information related to drought, especially in making decisions related to climate forecast information; hence, drought preparedness may suit them. In this respect, communities and farmers may be suspicious of a forecast and preparedness system if they do not understand or have some ownership of the scientific methods used to develop it, especially if they see the forecast and preparedness information as conflicting with their local traditional indicators (Patt and Gwata 2002). Chapter 12 gives a recent snapshot view of the use of such drought-related information.

While the value of forecasts and drought preparedness to rural communities and farmers will depend on their accuracy, it will also very much depend on the management options available to these key users to take advantage of forecasts (Nicholls 1991, 2000). Indeed, the value of climate and weather forecasting science related to drought to the end user, especially a farmer, may never have been demonstrated to the community by the institution developing and promoting the information. In Chapter 2, it was argued that trust in the scientific source was an important determinant of trust in the advice on risk and risk management. Trust in the source does not, however, ensure application. To the end user of drought modelling and information, the cost and benefits of different decision options are also a consideration (Katz and Murphy 1987; Gadgil et al. 2002; Hansen 2002). This aspect is reinforced by Sonka et al. (1987), who state that for benefits from scientific research in relation to drought issues to occur in farming practice it is necessary to identify those areas where tactical changes can be made either to take advantage of predicted (probabilistic) above-average rainfall or to reduce losses in predicted (probabilistic) below-average (drought) situations. In other words, climate and weather forecasts in relation to impending drought may have absolutely no value unless key management decisions have been identified through close interaction with farmers, and the farmers' key management decisions are capable of being changed by incorporation of climate and weather forecasts and information related to drought (Nicholls 1991, 2000; Hammer et al. 2001).

Bringing the end users into a triangle with policy makers and scientists does, however, bring other political problems. In Chapter 2 it was argued that the provision of more information does not always increase trust or adoption levels, since more information and engagement provide opportunities for additional disputation. In the workshop programmes, many of the farmers involved had attended 'climate and drought' workshops and so were well armed with information and understanding of the issues involved. Yet, with decreasing state income and the occasional lack of responses from federal agencies, many a department head was unable to respond to the endless calls for even more drought assistance, especially when nonrural-based political parties held power in government, and finance in this area was severely restricted. In 2002, one director general confided that the only way forward was for science to provide a 'way out' of this mess. Policy options had reached their capacity in being able to deliver to rural and other constituents. The lead author recalls meetings between government and rural lobby groups as being intensely stressful for all concerned. To this end the Queensland state government agreed to hosting a national 'Science for Drought' Conference in 2003, evidently the first in 16 years. At the outset, the aim was to bring scientists and policy makers together, whether they were state or federal people and to include international experts. To the lead author's dismay, a section head in government forbade the inclusion of drought policy makers in this national conference at the last minute; it was to be 'science only'. Nevertheless, a publication on 'science for drought' was forthcoming (Stone and Partridge 2003). One wonders about the potential richness of such a publication if it had, indeed, included both science and policy aspects in its content.

Nelson et al. (2007) may provide a way forward in managing the triangular relationship through the use of participative research together with the integration of relevant interdisciplinary research providing insights into the socioeconomic aspects of Australian drought policy (see Chapter 8). Indeed, socioeconomic data that also integrate such information as climate and agricultural production impacts can provide arguments to compare the value to decision makers of relatively imprecise, though extensive, information with relatively precise reductionist information and measures. In this approach Nelson et al. (2007) show that bioeconomic modelling can overcome the moral hazard and timing issues that have led to the dominance of historical rainfall and temperature data and analysis in the analytical support provided to Australian drought policy. This design also appears to overcome and isolate the production-related effects of climate variability on farm incomes from other causal variables such as changes in price and management techniques.

Furthermore, Meinke and Stone (2006, 221) argue that to overcome the disparity between climate/drought science and important policy aspects, the following is needed: '[a] participatory, cross-disciplinary research approach that brings together institutions (partnerships), disciplines (e.g., climate science, agricultural systems science, rural sociology and many other disciplines) and people (scientists, policy makers, and direct beneficiaries) as equal partners to reap the benefits from climate knowledge'.

They provide a useful diagram that shows the disciplines, relationships, and linkages for effective delivery of climate (and drought) information for decision making (Figure 7.1). Operational links are indicated by the solid arrows in the diagram and show connections that have already proven useful; dashed arrows indicate areas where operational connections still need to be better developed. They usefully point out that the core and basic principle of the concept (that is, the requirement of cross-disciplinary research

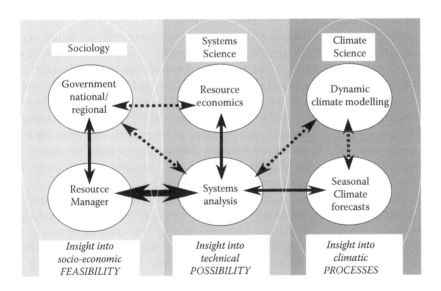

FIGURE 7.1
Key requirements for a systems and interdisciplinary approach: The aim is to convert insights gained into climatic and drought processes ('climate science') via systems analysis and modelling into the socioeconomic feasibility of decision options that would have direct relevance to improving drought policy needs. The approach is generic and can be applied by 'government national and regional' into an effective means of better integrating drought science and drought policy. (After Meinke, H., and R. C. Stone. 2005. *Climatic Change* 70 (1–2): 221–253.)

for effective delivery of scientific inputs) is generic and independent of the specific target discipline (from Meinke and Stone 2005).

Conclusions

As advocated elsewhere in this volume, there is much to be said for aspiring both to the inclusion of stakeholders in the science/policy interface and to broadening the definition of drought sciences. A key part of that would be translational scientists able to talk to the policy makers and end users, though these might need to be differently skilled and inclined people for each of the other points of the engagement triangle. This may, however, work better with a degree of independence for the scientists. As noted in Chapters 1 and 6 and illustrated in Chapters 10 and 11, many of the key US drought scientists and scientific agencies have a greater degree of independence than is the case amongst the key players in Australia, who are more likely to be located within the formal public sector. As shown in Chapter 6 and again in this chapter, this brings the scientists into institutions that can be changed very quickly.

The policy communities are especially difficult to work within because of the federal system, with tensions at both the intergovernmental political and bureaucratic levels. While state diversity in policy, science, and institutional arrangements provides an opportunity for comparative 'experiments', it also can be a barrier to cooperation and efficiency. Drought knows no administrative borders. Even without the complications of federalism, even single-level bureaucracies present a challenge. As shown here, institutional security and longevity can be limited and political masters can intervene on even the most seemingly innocuous activities. Policy makers want to control the interactions between the scientists and the end users, in case expectations are raised or inconvenient knowledge is revealed.

There are, however, also problems with the other components of the triangular relationship. Scientists hoping to use evidence to encourage drought preparedness, for example, may be frustrated by end-user lobbying of the policy makers that undermines some fundamental principles of a current policy framework. As an example, governments have been pushing for drought to be considered a natural event (Chapter 4), with a need for factoring this in to resources management, but at certain times have implicitly reverted to 'act of God' policies, with the accompanying agrarian rhetoric (Chapters 2 and 4). This suggests something of a dilemma for the scientist wishing to influence the policy process. Greater institutional independence may help to build trust with end users but these independent scientists may lose some of the direct line into the policy process.

References

Botterill, L. C., and A. Hindmoor. 2012. Turtles all the way down: Bounded rationality in an evidence-based age. *Policy Studies*. doi: 10.1080/0144-2872.2011.626315.

Choi, B. C. K., T. Pang, V. Lin, P. Puska, G. Sherman, M. Goddard, M. J. Ackland, P. Sainsbury, S. Stachenko, H. Morrison, and C. Clottey. 2005. Can scientists and policy makers work together? *Journal of Epidemiology and Community Health* 59:632–637.

CLIMAG. 2001. *Newsletter of the Climate Variability in Agriculture R&D Program.*

Gadgil, S., P. R. Seshagiri Rao, and K. Narahari. 2002. Use of climate information for farm-level decision making: Rainfed groundnut in southern India. *Agricultural Systems* 74 (3): 431–457.

Hammer, G. L., J. W. Hansen, J. G. Phillips, J. W. Mjelde, H. Hill, A. Love, and A. Potgieter. 2001. Advances in application of climate prediction in agriculture. *Agricultural Systems* 70 (2–3): 515–553.

Hansen, J. W. 2002. Applying seasonal climate prediction to agricultural production. *Agricultural Systems* 74 (3): 305–307.

Katz, R., and A. H. Murphy. 1987. Forecast value: Prototype decision-making models. In *Economic value of weather and climate forecasts*, ed. R. Katz and A. H. Murphy, 183–218. Cambridge: Cambridge University Press.

Lofstedt, R. 2005. *Risk management in post-trust societies*. Houndmills: Palgrave Macmillan.

Lowe, I. 2002. *The need for environment literacy*. ALNARC Online Forum www.staff.vu.edu.au/alnarc/onlineforum

Ludwig, D. 2001. The era of management is over. *Ecosystems* 4:758–764.

McLain, R. J., and R. G. Lee. 1996. Adaptive management: Promises and pitfalls. *Environmental Management* 20 (4): 437–448.

Meinke, H., and R. C. Stone. 2005. Seasonal and inter-annual climate forecasting: The new tool for increasing preparedness to climate variability and change in agricultural planning and operations. *Climatic Change* 70 (1–2): 221–253.

Meyer, M. 2010. The rise of the knowledge broker. *Science Communication* 30:118–127.

Nelson, R., P. Kokic and H. Meinke. 2007. From rainfall to farm incomes—transforming advice for Australian drought policy. II Forecasting farm incomes. *Australian Journal of Agricultural Research* 58: 1004–1012.

Nicholls, N. 1991. Advances in long-term weather forecasting. In *Climatic risk in crop production: Models and management in the semi-arid tropics and subtropics*, ed. R. C. Muchow and J. A. Bellamy, 427–444. Wallingford: CAB International.

———. 2000. Opportunities to improve the use of seasonal climate forecasts. In *Applications of seasonal climate forecasting in agricultural and natural ecosystems: The Australian experience*, ed. G. L. Hammer, N. Nicholls, and C. Mitchell, 309–327. Dordrecht, The Netherlands: Kluwer Academic Publishers.

Parsons, W. 2002. From muddling through to muddling up—Evidence based policy making and the modernisation of British government. *Public Policy and Administration* 17:43–60.

Patt, A., and C. Gwata. 2002. Effective seasonal climate forecast applications: Examining constraints for subsistence farmers in Zimbabwe. *Global Environmental Change* 12:185–195.

Queensland government. 1992. *Drought: Managing for self reliance: A policy paper.* Brisbane: Queensland Government.

Sonka, S. T., J. W. Mjelde, P. J. Lamb, S. E. Hollinger, and B. L. Dixon. 1987. Valuing climate forecast information. *Journal of Climatology and Applied Meteorology* 26:1080–1091.

Stone, R. C., and H. Meinke. 2007. *Weather, climate, and farmers: An overview: Commission for Agricultural Meteorology Special Report.* Meteorological applications. Geneva United Nations' World Meteorological Organisation.

Stone, R. C., and I. Partridge, eds. 2003. *Science for Drought: Proceedings of the National Drought Forum.* Brisbane: Queensland Department of Primary Industries.

8

Reflections on Evidence-to-Policy Processes

Daniela Stehlik

CONTENTS

As noted in Chapter 1, there has been something of a trend, or at least an increasing aspiration, to develop policy based on evidence that the proposed action will achieve its intended outcomes. The evidence can come from other jurisdictions, extrapolations from studies of other policy areas, clinical and field studies, and so on.

Public policy is also understood to involve *consultation* with those stakeholders for whom the policy is being developed, or those potentially impacted by it, either positively or negatively. As discussed in Chapter 2, there is some tension between an evidence-based approach, which tends to privilege the 'experts', and consultation that fits with a democratic approach but has its own difficulties. This tension could be resolved to some extent if a consultation process is treated as another means of gathering evidence—for example in relation to the extent and impacts of the problem and the acceptability of particular policy solutions. Such evidence from 'stakeholders' is then gathered and becomes part of the evidence base that informs the policy process and part of the *proof* needed to substantiate the manner in which any subsequent policy framework is developed. This proof is tacitly promoted as being as rigorously scientific as possible. By scientific here is meant that it offers *certainty* or at least authoritative advice both to the policy maker and to the politician for whom the evidence has been gathered. There are, however, reasons examined in this case study as to why feedback from those affected by particular phenomena, such as drought, might be considered less

117

worthy evidence than that from other sources. Furthermore, the 'evidence' gathered from those most affected by natural or social forces can be seen as tainted, because there is both self-interest and emotionalism at play. This is especially the case if the consultation is undertaken during a period of high stress, as in this case study.

In a major address in 2009, Gary Banks, commissioner of the Productivity Commission,* pointed to 'the challenges of properly implementing an evidence-based approach to public policy—and of being *seen* to have done so, which can be crucial to community acceptance of consequent policy decisions' (Banks 2009, 2). In a similar vein, Cockfield also notes in Chapter 2 that some of those involved in risk management research argue for the need to build confidence in decision-making and advisory bodies. There are therefore some practical and ethical arguments for public visibility in the evidence-to-policy process, but despite this visibility, there is no guarantee that it will achieve community acceptance, be interpreted into policy, or even if it is, that the politicians will adopt such evidence. Regardless of this lack of guaranteed inclusion or consideration, however, the consultation process itself remains assumed to be both objective and vital.

Banks suggests that an evidence-based approach enables political judgements themselves to be more objective. Banks (2009, 4) argues that 'policy decisions will typically be influenced by much more than objective evidence, or rational analysis. Values, interests, personalities, timing, circumstance and happenstance—in short, democracy—determine what actually happens'. He proposes:

> Without evidence, policy makers must fall back on intuition, ideology, or conventional wisdom—or, at best, theory alone. And many policy decisions have indeed been made in those ways. But the resulting policies can go seriously astray, given the complexities and interdependencies in our society and economy, and the unpredictability of people's reactions to change.

Those of a similar rationalistic turn of mind might be especially concerned about policy development during crises, such as droughts, where there is pressure for action, emotions may be running high, and people's core values or fears may be to the front of their thinking.

Hence, consultation with stakeholders to begin to draw together some evidence of impacts, often undertaken in the midst of a crisis for which the policy is being demanded, tends *not* to be viewed as reliable as evidence obtained 'scientifically'. In other words, the reality is that such consulted views remain presented as essentially personal (local, subjective), and these are then weighed against the amassed (objective) 'scientifically proven data'

---

* The Australian Productivity Commission 'is the Australian government's independent research and advisory body on a range of economic, social and environmental issues affecting the welfare of Australians' (http://www.pc.gov.au/ retrieved 15 March, 2012).

and often found wanting. This process of weighing up the evidence is, however, at best, itself a subjective political balancing act. Despite Banks's call for rigour, all actors attempting this balancing act filter such evidence through their own frames of reference. In this process, some evidence becomes foregrounded while other evidence retreats, or even disappears. However, any methods associated with this process of filtering (or perhaps more kindly termed *refinement*) are rarely discussed and therefore such values that underpin this process are never made overtly transparent.

This chapter suggests, based on the case study discussed here, that the process by which evidence gathered during consultation is rationally and logically transformed into policy can therefore be viewed as essentially one of illusion. While it appears to be relatively straightforward, it remains both complex and opaque. For Banks, such evidence-to-policy processes also inform policy makers' judgements in that 'they can also condition the political environment in which those judgements need to be made' (Banks 2009, 4). In other words, even the process of gathering such evidence, as well as its refinement, is, at heart, essentially political and therefore subjective rather than objective.

This chapter aims to explore this balancing act to decision making in more detail, within the political context in which it is undertaken, through the personal experience of the author, who was a member of a national review established in 2008 to enable the development of a new federal drought policy framework. This included a major piece of community consultation and this process and its subsequent transformation into policy is discussed later. As the review process was focussed on social issues and impacts and it drew on the social sciences, it is useful to consider this aspect of the context first.

Striving for Certainty?

To the outsider, the assumptions and the processes through which decision makers decide which evidence to consider, how it becomes refined, and how it then becomes integrated into policy remain essentially ambiguous. Consultation with affected individuals and communities has two roles in the policy process. First, qualitative social scientific research uses methods that can broadly be described as 'consultation' in the investigation of specific research questions and the conclusions of that research effort can then be an input into policy. These methods range from relatively well understood approaches, such as one-to-one interviews, focus discussions, town hall meetings, or community forums, to more complex and intensive approaches such as deliberative democracy.* Second, consultations can be an explicit part

---

* For more detail, see, for example, Gastil and Levine (2005).

of the policy development process and are set up for that purpose. These can include major reviews (discussed further later) and community consultations at the local, regional, or national levels. The social science input into the 2008 drought policy review included both of these types of research: the outcomes of research conducted independently of the review and the results of consultations specifically set up for the policy process. As Bridgman and Davis (2000, 76) note, 'The legitimacy of much public policy now rests on an exchange between citizens and their government. Public servants and politicians must find ways to discuss with relevant communities of interest and draw them into the policy process…'.

In spite of these two avenues for incorporating social issues in the policy process, it has been the experience of many social scientists that their contributions appear to be dismissed subsequently in the policy process as being merely 'anecdotal' when rich qualitative evidence is gathered—a debate that has now been at the heart of research methodology for over a century.

To discuss this as relevant to this chapter, I am drawn to the recent writings of two of Australia's premiers*—both social scientists and both reflecting after their time in public office—in raising the issue of evidence to policy. Both were premiers of the same state, Western Australia, then and now a major resource state in the nation. Both were also Labor (social democrat) premiers, and as reformist leaders, they would likely have agreed with the statement by former Labor Prime Minister Kevin Rudd, that 'evidence-based policy making is at the heart of being a reformist government' (cited in Banks 2009, 3).

For Carmen Lawrence,[†] while the social sciences *should* also be 'at the heart of good public policy,…they are not'. She suggests that this is primarily because 'many policy-makers assume they know, because they are human and live in society, all that they need to know about human society and behaviour' (Lawrence 2010). Lawrence suggests that 'while nearly all public policy rests on assumptions about human behaviour, these assumptions are rarely made explicit or tested against the available evidence. And sometimes they are simply wrong'. She argues that the best way to integrate the social sciences into good policy making is to 'build close collaborative relationships between knowledge users and research providers' as this will then enable each to 'better understand the other's domain'.

---

* A premier is the leader of the party in government at the state level in Australia's federal system.
† Premier of Western Australia 12 February 1990 to 8 February, 1993. Federal member (1994–2002) and federal minister of human services and health. PhD. Psychologist. Currently Winthrop Professorial Fellow at the University of Western Australia. (http://john.curtin.edu.au/lawrence/biography.html. Retrieved 15 March, 2012).

Geoff Gallop has another view of the social sciences and their place in policy making. For Gallop,* the social sciences are critical when partnered with social activism in 'developing schemes for human improvement' as they 'have the potential to assist us in finding answers to questions...[about] different ways of organizing our common affairs' (Gallop 2007, 23). He remains concerned that the critical edge offered by the social science disciplines is in danger of being lost, perhaps (although he is not explicit) because of the very collaboration argued for by Lawrence, as this has tended to become more of an unstated but agreed to symbiosis between parties in their ongoing search for evidentiary certainty on which to base policy decisions.

Gallop cautions against the tendency to offer (social science) certainty to policy makers and politicians, as there is no such thing as a 'standard citizen', and policies must allow for society's diversity. He also warns against the potential for a loss of the 'critical and predictive value' of the social sciences when consideration of difference is ignored. His warning highlights the need to ensure that 'scientific rigour' does not become confused with a tendency to offer 'guaranteed success'. He concludes that while the social sciences 'cannot tell us what to believe,...they can inform and temper our beliefs' (2007, 23).

Both commentators highlight how the place of the social sciences within policy decision making (and therefore within the evidence-to-policy process) continues to remain, at best, marginal. Lawrence (2010, 32) suggests that the tendency for the rejection of critical social science evidence is because not only do 'few politicians in Australia have tertiary level science training [but also] more than a few have only a peripheral interest in policy development'.

In summary, it would therefore be reasonable to conclude that the arena in which the delicate and political balancing act of evidence to policy is played out is one where misunderstandings and misinterpretations can and do exist. The chapter now turns to explore this arena in more detail, from a personal perspective, drawing on recent and relevant experience.

A Tripartite Review

Early in 2008, the newly elected federal Labor government announced a series of national consultations that culminated with an event in Canberra called *Australia 2020*. The Labor Party and especially its leader, Kevin Rudd, were interested in developing a reform agenda, perhaps following from the previous Labor governments (1983–1996), which were notable for major economic

---

* Premier of Western Australia 10 February 2001 to 16 January 2006. PhD. Economist. Currently professor and director of the Graduate School of Government at Sydney University (http://john.curtin.edu.au/gallop/biography.html. Retrieved 15 March, 2012).

reforms. As noted earlier, Rudd sees evidence-based policy as a foundation for 'reformist' governments. One thousand Australians from all sectors were invited, and deliberations were held on future national issues of importance over a weekend. This event was presented as being an opportunity to 'capture ideas—both large and small—that can be shaped into concrete policy actions' (Commonwealth of Australia 2008a, 1). Among the many issues raised in various forums were those of climate change and the impact of the ongoing (at that time) drought on rural Australia—specifically, the 'reform of current drought assistance welfare' (Commonwealth of Australia 2008b, 19).

This call for reform had become a major theme for the newly elected Labor government, as the drought at that time had already covered nearly 90 percent of some states and was then in its fifth year in others. In March 2008, the prime minister highlighted the need to review the policies underpinning the responses to drought. He stated that the current drought policy was 'based on a model of a one-in-25 year drought and assumes rainfall will return to past seasonal conditions and does not factor in climate change. This approach needs updating' (reported in Maiden 2008).

As the millennium drought took hold across Australia, it became clear that previously assumed time lines between drought events were being overturned, and the 'drought of the century' (the mid-1990s drought) was being followed almost immediately by one as severe, if not more severe, and widespread.

For policy makers this created a real and pressing challenge as the policies developed in the mid to late 1990s and from earlier policy reviews were being put to the test, in practice, immediately. There was not the 'luxury' of a decadal 'gap' of time either for the countryside to recover or to integrate new or amended policy into more deliberate programme planning. The decades of policy leading up to the millennium drought and the reason it needed urgent policy review was that it was based on a key statement in the previous policy platform of 1989 to the effect that drought represented 'a prolonged failure or inability of *producers* to respond to...deteriorating conditions' (Drought Policy Review Task Force 1990, vol. 1, 7, italics added). In other words, drought was for producers, farmers, and land managers to manage.

The drought of the new millennium challenged this previously agreed policy platform by moving 'beyond the farm gate' and entering the cities of the eastern seaboard of Australia. This was in stark contrast to the 1990s drought, generally considered then as one of the worst in Australia's history, which had *not* been framed as a city issue. As a result, for the government, the lack of water for cities highlighted that this drought event was much more than simply a 'farming problem'. It had become a serious *political* challenge for federal policy, which had previously largely viewed drought as an issue solely for agriculture and therefore the responsibility of one particular minister and one particular government agency. This is stated as evident in the 'contextual overview' of the Report of the Social Impacts Review, which highlighted the transition of Australia from a 'rural society to an urban' one and the changes to policy associated with the forces of urban drift, global free markets, and economic structural change (Kenny et al. 2008, 6).

So the political environment in late 2007/early 2008 had put pressure on the newly elected Labour government to *do something** and as the impacts of the drought became increasingly caught up in the larger discussion of global warming and climate change, the government responded by proposing a tripartite review process focussing on the three aspects of the policy problem: likely future climate patterns, the economic effectiveness of existing policy settings, and the social impact of drought. In spite of the broader context, the review remained largely confined to considerations of the impact of drought on agricultural producers and rural communities, consistent with earlier policy approaches.

The climate component was directed to the Bureau of Meteorology (BOM) and the CSIRO, which were charged with examining 'the implications of future climate change for the current exceptional circumstances (EC) standard of a one in 20–25 year event' (Hennessy et al. 2008, 1). The economic component was directed to the Productivity Commission, which was asked to report on the 'appropriateness, effectiveness and efficiency' of the various government support measures in place to manage drought (Australia. Productivity Commission 2009, v). For the Bureau of Meteorology, the science associated with the modelling of climate variability had already been well established (see Chapter 3 in this book) and their report was undertaken in a short period of time by drawing on various sources of previously published data (including from CSIRO). For the Productivity Commission, their economic analysis rested on a review of the taxation arrangements as well as the many varied and complex federal/state arrangements associated with income support, subsidies, and assistance provided to sustain national agricultural production. The Productivity Commission undertook a national consultation process with hearings in all capital cities except Hobart and Darwin and in five regional centres, and it received 192 submissions. The delivery of its report was timed to be able to take into account the findings from the social component of the review.

All three components of the tripartite review were stated as contributing to a more certain approach to the decision making for policy development after the review itself, and it was intended that this would directly inform the 2009–2010 federal budget process. For example, when accepting the Productivity Commission's report, the federal minister stated that 'there is widespread recognition that the current system is unfair, *creates uncertainty* and doesn't allow for assistance to be provided until farmers reach crisis point. We need to provide greater support to rural communities and build a strong future for rural Australia' (Burke 2009, italics added).

---

* The IPCC released its 'Impacts, Adaptation and Vulnerability Report' in the first week of April 2007, and media interest heightened subsequently (see, for example, Dayton 2007).

The 2008 review was an important departure from previous drought policy development processes. The last time a major independent review of drought policy had been conducted was by the Drought Policy Review Task Force, which reported in 1990. The 2008 review was notable because it explicitly included consideration of the social impact of drought and the long-term implications of climate change for drought occurrence. Previous reviews had been framed within concerns about structural adjustment in agriculture. However, the inclusion of social issues did not guarantee that they would be given equal weight with the evidence emanating from the other components of the review.

The Evidence Gathering Process

The third and final component of the review comprised both a significant consultation and evidence-gathering process, focussed on the social impacts of drought on farm families and rural communities and an analysis of existing social scientific research on drought impacts. This aspect of the review attempted to integrate the local experiences of people living with drought across the country with 'independent research' (Kenny et al. 2008, iii). In this sense, it is an example of the drawing together of social scientific and local knowledge into the evidence-to-policy process and offers an opportunity to examine this in more detail.

Figure 8.1 outlines the various sources of evidence as gathered through both consultations and commissioned research for the social component of the review.

The composition of the Expert Social Panel (the Panel) created to undertake this element of the review could also be seen to personify examples of local and science knowledge and experience. While increasingly a more popular approach within the political system of data collection, consultation processes can encounter their own legitimacy problems either because they are seen as 'political' appointments and are therefore naturally 'biased' in their outlook and reporting, or because they become dominated by self-selecting and potentially unrepresentative spokespersons (Bridgman and Davis 2000, 77). There remains a scepticism about their capacity to pull together complex information in the often short time they are given. Finally, there is a tendency to consider them more as 'political stopgaps' to show that something *is* being done, without any real decisions having to be taken.

The membership of seven (three men and four women) on the Expert Social Panel between them represented farming, grazing, horticulture, health policy and practice, education, social policy; nongovernment expertise, previous members of federal Parliament, and agribusiness as well as graduate

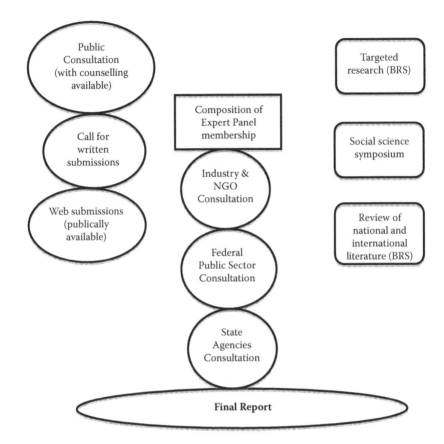

FIGURE 8.1
Schematic outlining the sources of evidence gathered during the expert social review of drought policy, June–October 2008.

and postgraduate experience across a variety of disciplines. There was state representation as well, as all (except Victoria) were represented.

The Panel was supported by a small team from within the federal government Department of Agriculture, Fisheries and Forestry (DAFF) and from within Centrelink* by a team of social workers who joined the Panel at its various forums across the country. The support team also managed expectations from other federal agencies, as well as any media interest raised during the Panel's progress.

It was always appreciated that there would be a short time line both for consultation and report writing (the Panel had 12 weeks in total), and this in turn determined the methods adopted to gather the evidence. It was also understood that there needed to be public forums of some kind, due to the concern about the impact the drought was having and would continue

---

* Centrelink is the major federal agency that provides both income support and counselling.

to have (there being no evidence of its breaking during 2008) on people's well-being.

In this sense, it was very much a public performance, as much as it was a process of gathering evidence, and was a way of demonstrating the government's concern about the diminishing quality of life people were experiencing. The Panel understood this aspect of 'review as performance', and the public forums were staged in major venues in key towns, advertised widely (advertisements were placed in 21 newspapers), and well attended. These are summarised in Table 8.1.

In total the 25 forums drew some 670 participants. The Panel travelled the length and breadth of the country and visited each state. All the sites were chosen because they were located in places that had been drought declared. In some cases—for example, in central western Queensland—the wetter seasons between the drought of the 1990s and this current drought were very few indeed and the Panel heard of experiences drawn from almost two decades of continuous drought.

In addition to this public consultation, the Panel also consulted a number of key organisations. In some cases, this was in response to a request by that organisation; in other cases, the Panel sought advice. Table 8.2 details these by sector.

The social impact component of the review established a dedicated website as part of the consultation process. It invited written submissions, either directly to the website or by mail. These submissions were primarily signed by individuals, in the case of two on the website they preferred to remain anonymous, and if permission was granted, were loaded onto the public website during the period of the review (and still remain there as an archive). Over 230 submissions were received. In some cases, submissions were received from individuals who could not attend a public forum. In other cases, people attended the public forum and then wrote a subsequent submission. Some of the organisations/agencies listed before also presented written submissions, sometimes of great detail. A summary of the submissions by type is presented in Table 8.3.

Two further opportunities for specific consultation were undertaken by the Panel during the review. The first was a workshop held in Canberra in July 2008 to which research and education experts (primarily but not exclusively from universities) were invited. A facilitated discussion was then held to draw out the key issues associated with current social science research as it related to the terms of reference for the Expert Social Panel. A further consultation was also arranged with invited staff from the key federal government agencies involved in drought support. These senior staff met with the Panel in Canberra and discussion was wide ranging and free flowing, rather than specifically focussed in the way in which the public forums had been.

As Figure 8.1 highlights, the consultation process was both broad and detailed. It created an important baseline understanding across sectors, across states, and across farming properties and communities of the local

TABLE 8.1

Public Consultations Undertaken by Expert Social Panel (July–August 2008)

Date	State	Location	Venue	Participants
21 July	Northern Territory	Alice Springs	Crowne Plaza	7
24 July	New South Wales (NSW)	Inverell	RSL Club	45
25 July	NSW	Bourke	Bourke Bowling Club	43
18 August	NSW	Gilgandra	Gilgandra Services Club	50
19 August	NSW	Forbes	Forbes Service Memorial Club	50
21 August	NSW	Griffith	Catholic Club Yoogali	60
22 August	NSW	Goulburn	Goulburn Workers Club	75
28 July	Western Australia (WA)	Esperance	Esperance Bay Yacht Club	15
29 July	WA	Morowa	Morowa Town Hall	32
30 July	WA	Wongan Hills	Wongan Hills Hotel	18
30 July	WA	Merriden	Merriden Regional Leisure Centre	28
4 August	Victoria	Shepparton	Shepparton Club	99
5 August	Victoria	Birchip	Birchip Leisure Centre	75
25 August	Victoria	Colac	Colac Bowling Club	34
26 August	Victoria	Mildura	Mildura Settlers Club	37
6 August	Tasmania	Bothwell	Castle Hotel	53
11 August	Queensland (Qld)	Gatton	Gatton Bowling Club	20
11 August	Qld	Dalby	Dalby RSL Club	25
12 August	Qld	Charleville	Racecourse Complex	30
13 August	Qld	Longreach	Longreach Civic Centre	27
13 August	Qld	Emerald	Emerald Town Hall	40
25 August	South Australia (SA)	Keith	Keith Football Club	55
26 August	SA	Gawler	Gawler Arms Hotel	46
27 August	SA	Orroroo	Blacksmith's Shed	37
27 August	SA	Wudinna	Wudinna Community Club	39

Source: Derived from Kenny, P. et al., 2008, *It's about people: Changing perspectives on dryness—A report to government by an expert social panel.* Canberra: Commonwealth of Australia, pp. 99–100.

TABLE 8.2

Consultations with Individuals and
Organisations by Sector (July–August 2008)

Sector Represented	Number
Local government	1
Finance institution	1
Land Councils	2
Health organisations	3
Welfare NGOs	4
National representative organisations	4
R&D institutions	5
Mental health organisations	5
Federal government agencies	6
Agribusiness/agri-industry	6
State government agencies	8
State representative organisations	8

Source: Derived from Kenny, P. et al., 2008, *It's
about people: Changing perspectives on dryness—A report to government by an expert
social panel.* Canberra: Commonwealth of
Australia, pp. 101–102.

knowledge associated with social impacts. Kloppenburg (1991, 537) reminds
us (as the Panel was to discover) that such local knowledge reflects 'the sense
of inseparability from a particular place', and while there were many differences in impact, there were also some important similarities, and this
enabled the emergence of a number of common themes.

A further aspect of the gathering of evidence was focussed on ensuring
that the Panel was up to date with existing (social science) national and international research in the area, as well as in its consideration of the potential
of commissioning specific research to assist in its deliberations. This aspect
was as a direct result of early discussions within the Panel of the need to
have a balance established between quantitative evidence and qualitative
evidence. This task was undertaken by research scientists from the social
sciences programme within the Bureau of Rural Sciences and the Panel subsequently commissioned the following activities:

- An analysis of the social circumstances of rural people and communities drawn from the HILDA[*] data set

- An analysis of the social circumstances of farmers from the June
  2008 BRS social sciences programme climate change and industry
  adaptation survey

---

[*] Household, Income and Labour Dynamics in Australia Survey. See http://melbourneinsti-
tute.com/hilda/ for further details.

TABLE 8.3

Submissions to Review by Type

Sector Represented	Number
Individual Submissions	
Queensland	9
New South Wales	31
Victoria	24
Tasmania	5
Northern Territory	1
South Australia	17
Western Australia	5
Unknown location	7
Churches	1
Finance institutions	1
National health organisations	1
National mental health organisations	1
Federal government agencies	1
Members of Parliament (state/federal)	2
Land management	2
Regional Landcare organisations	3
Service clubs	3
National welfare NGO	4
National representative organisations	5
R&D/education institutions	5
Agribusiness/agri-industry	5
State government agencies	8
State health and mental health organisations	10
State welfare NGOs	20
State representative organisations	21
Local government	23

Source: Derived from Kenny, P. et al., 2008, *It's about people: Changing perspectives on dryness—A report to government by an expert social panel.* Canberra: Commonwealth of Australia, pp. 103–107.

- An analysis of a quality of life survey of farmers and farm workers in drought-affected areas utilising the Australian Unity Wellbeing Index[*]
- A literature review (Aslin and Russell 2008) revealing a range of concepts and frameworks to help understand the impact of drought and how rural people and communities respond to it

The Panel invited presentations of the findings from the research in early August.

---

[*] Based at the Australian Centre on Quality of Life at Deakin University; see http://www.deakin.edu.au/research/acqol/auwbi/index.php for further details.

All formal consultations concluded at the end of August 2008 (see Table 8.2) and the majority of the commissioned research was presented to the Panel by early to mid-September 2008 for consideration. The Panel members were working to a very tight deadline, marked by two previously agreed upon dates. The first, a presentation to the Primary Industries Ministerial Council (PIMC), was to be held in Canberra on 19 September. The second was the launch of the report by the Federal minister, scheduled for 30 September. This left the Panel with little time (less than 2 weeks) to absorb, reflect, and discuss the evidence it had drawn together. The next section deals with the process of refinement as it occurred throughout the review itself, offers an example of such refinement, and then concludes with some reflections.

Refinement

This section considers the filtering of evidence in the transition to policy and the aspects associated with the process itself within the case study being discussed. The process of refinement can be considered as a series of concentric circles, each bordering or edging the evidence toward a core series of common themes or elements. This approach suggests a logic and order to the process, with deep discussion and, if necessary, consensus among the participants reached, rather than what actually became quite a dynamic, messy, and, at times, passionate experience.

If we do consider these concentric 'filters' in this way, it can be seen that the terms of reference of the review offered the first 'boundaries' to the refinement process, in the sense that they defined what was 'within the process' and what should be excluded from it. The terms of reference also determined the consultation process itself (full details of these can be found in Appendix 2 in Kenny et al. 2008, 92–94). A further refinement occurred as the places for community forums were identified and agreed to.

Most obviously, the refinement process was directly influenced by the choice of membership of the Review Panel itself. As detailed earlier, this group of seven members represented a variety of aspects of the issues under review, drawn from different sectors and coming to the issues from their own particular view (stand) points. For those on the Panel who were themselves active farmers or land managers, the review offered an opportunity to draw on their experience and consider the changes and adaptations underway as a result of the impact of drought on farm practice. For those on the Panel whose primary interest was the social impacts, the review consultations provided opportunities to explore these in more detail, and to attempt to understand how change was underway within family structures and community cohesion. It should be recalled that the Panel also consisted of members from agripolitics and from federal political arenas. These were also the same people who were farmers/graziers or land managers, so their consideration of the evidence also then became politically framed. For those Panel members who had spent time in policy development and policy planning, the evidence was drawn into an analysis of the current policy framework and the potential and possibilities associated with new policy development.

After each community forum, the Panel members took time to 'debrief' within the group, and begin to 'member test' the evidence with each other as it was being presented to them. Drawing on their personal experiences, individual Panel members began to identify the key issues that were emerging for them. As the consultation process was an intense one over a very short period of time (see Table 8.1), this 'debriefing process' was largely undertaken 'on the run'—that is, during the review consultation process itself—in the minibus, on planes, in cars, and around meals. This informal process enabled the group to trust each other, and as that trust developed, the discussion became more candid in breadth and richer in depth.

Such discussions as were held were done so with the participation of the small team of staff from the Department of Agriculture, Fisheries and Forestry. These individuals had been selected to assist the Panel because of their experience in the department and their understanding of the current agricultural policy sector more broadly. Their primary task was to ensure that the terms of reference remained adhered to, and that the time frame given was not breached. Drawing on their own frames of understanding, these individuals also influenced the refinement process, sometimes overtly, sometimes more subtly. It needs noting that all such discussion was undertaken frankly and openly. In this sense, the integration of science and local evidence as undertaken in the review does follow Lawrence's argument for best practice being based on 'close collaborative relationships' (Lawrence 2010, 32).

It may be useful at this point to consider one aspect of a theme that emerged from the process for the Panel and became a topic of much ongoing discussion, and one aspect of it then became part of the final report. This issue also re-emerged later, following the policy framework development of a pilot programme for drought relief that was undertaken following the government's acceptance of the review documents. A further aspect of this issue and the consideration of the Panel to it did not make the final 'cut', and the matter therefore offers an interesting case example of the filtering process. The issue itself was the vexed question of the place of rural Australia in the nation.

The 'Value' of Rural

The extensive and ongoing nature of the millennium drought had created an environment of public concern that became very obvious to the Panel through a regular and consistent theme that could be summarised informally as *the drought has meant that the rest of the nation has forgotten us* or *we are doing it so tough, so why doesn't Australia recognise our value to the nation?* It was a topic regularly stated in various ways in the forums, and it also emerged as a recurrent theme within many of the written submissions. Despite the millennium drought hitting the cities hard, rural people still felt alone in facing its consequences and still, to a certain extent, felt blamed for the problem. See Chapters 2 and 4 for more discussion of agrarianism, one element of which is a perceived urban/rural divide.

Some snapshots of the comments heard by the Panel at forums and read in submissions will give the overall 'flavour' of the issue as presented:

> *You'd have to wonder if the government really wants to help. The amount of regulation [is] overwhelming. [Does the government] want primary production in Australia?* (local government spokesperson. Community Forum. Central Western Queensland, August 2008)

> *The drought makes us feel like we are asking for something we are not worthy of…* (female farmer, Community Forum, Tasmania, August 2008)

> *[Let's]…keep what we have that is good and worthwhile in this country [that] has occurred [as a result of the] hard work of many generations of Australians who have [made] this country what it is today* (submission no 218. Mildura, Victoria)

> *A lot of farmers ask themselves, 'Why am I getting up in the morning?' We have to value the farmers…* (community forum, Central Queensland, August 2008)

> *No one is talking about a future…* (community forum, Central New South Wales, August 2008)

In response to this evidence, the debate within the Panel became one of how best to reinterpret this concern to government so that the immediacy of it and the consequent longer term impact of it on individual health and wellbeing (Stehlik 2009) could be best understood. The Panel understood that while any future drought policy needs to be consistent with the long-term vision for the nation, the issue was not one of either money or services, so how could government best respond?

Two suggestions related to this theme emerged from the informal Panel discussions. The first was that a major statement could be made, expressly by the prime minister to give it the necessary status, as to the value to the nation of those Australians who contributed its food and fibre. It was imagined that such a statement would be made in Parliament, alongside similar previous statements on important national matters and would then be widely publicised* (with hope for a bipartisan approach) and thereby give

---

* The Panel was influenced by the recent major statement made by the prime minister in federal Parliament on 13 February 2008—*An Apology to the Stolen Generations* (see http://www. abc.net.au/news/2008-02-13/rudd-we-say-sorry/1040976. Retrieved 2 April, 2012).

the impression that rural Australia had not been forgotten. It was considered by the Panel that such a statement could also include reference to the future drying of the nation's climate, and that responses to this future were a 'whole of nation' issue, rather than simply the responsibility of those on the land. Such a statement would also be an immediate response to the review process and would send the important message that people's concerns had been heard and heard loudly.

The second idea was aimed more at a longer term response (see Chapter 3), with an eye on the future, which, as was clear from the scientific evidence, would be an increasingly drying and challenging one. This was a suggestion that Australia prepare something akin to *The Cork Declaration—A Living Countryside* as established by the European Union (EU) in 1996 The EU Declaration states the intrinsic value to the member states of the European countryside and proposes a ten-point plan for its future rural development. An Australian declaration similar to this was envisaged in the Panel's informal discussions, which would then frame any future policies that affected rural Australia, regardless of whether they emerged from the agricultural agencies, or the health, welfare, or infrastructure agencies. The Panel considered that such a declaration could also assist in framing any future policies for Australia's drying future.

Both ideas were considered for some time by various members of the Panel; however, in the final process of refinement, which occurred over two day-long meetings, where all the evidence was being weighed up and the final report drafted, only one of these suggestions (the first) made the final 'cut' and this was in a format slightly different from that envisaged by the Panel. It emerged as Recommendation 2 of the Social Impact Report (Kenny et al. 2008, 21, 67), which stated that 'governments must make a high-level statement of commitment to a strong, healthy, vibrant and sustainable rural Australia'.

This statement eventuated as not being made by the prime minister in an immediate response to the drought, but rather as a statement from the Minister for Agriculture, Hon Tony Burke, some 20 months later, when delivering his Second Reading Speech for *The Farm Household Support Amendment (Ancillary Benefits) Bill 2010,* which made certain legislative changes for drought support, by including in this speech that 'the Rudd government believes in the strength, innovation and resilience of our rural Australia' (Burke 2010, 4118).

This did recognise Recommendation 2 as proposed by the Expert Panel, but however interesting Second Reading Speeches[*] may be to those of us involved in policy analysis, they are not generally read by the broader population, and in this case, the speech itself was not widely reported, nor was this statement of support widely promulgated. It therefore did not have the necessary and immediate impact as imagined by the Panel in its informal

---

[*] A Second Reading Speech is given by Ministers when legislation is introduced into Parliament and is a formal statement of the government's intent with regard to that piece of legislation.

discussions leading up to the recommendation in its report. There were, however, two later notably supportive statements. Following the 2010 federal election, the Labor Party required the support of two rural independents to form government and, in an agreement with them, the importance of rural people and industries was emphasised. Then again the same sentiment was restated, although slightly differently, within the principles for the Drought Policy Reform as stated by the Primary Industries Ministerial Council in April 2011 as: 'Principle 3—recognition of the important role of farmers as the nation's food producers' (cited in Keogh and Granger 2011, 8).

At the time of writing, the Australian governments have yet to decide on the components of a new drought policy framework. It is therefore not clear how much relative influence the three reviews will have on the policy process. However, a drought policy 'pilot' run in the state of Western Australia from 2010 may provide some hints as to the shape of the new policy. The components of the pilot were a mix of social support and structural adjustment initiatives. A review of the pilot appeared to prioritise social concerns in its recommendation that 'the following measures would represent a robust future policy platform':

- An income support safety net for farm families in hardship that is available based on demonstrated individual need
- The permanent presence of social support services delivered via outreach to people in rural communities
- Continuing opportunities to engage in and implement strategic farm business planning
- Ongoing access to the FMD [Farm Management Deposits] scheme and existing tax incentives for primary producers (Keogh and Granger 2011, 3)

Reflections

It has been suggested that consultation to develop evidence for policy is 'predicated on an acceptance by policy makers that those being consulted have the capacity not only to comment, but to influence the final disposition of the policy proposal' (Bishop and Davis 2002, 22). In the case of the review of drought policy for Australia in 2008/2009 there was an acceptance that, as the current policy had run its course and as droughts appeared to be recurring with much more frequency, it was urgent to develop new policies based on the lived experiences of land managers, farmers, and rural communities.

The tripartite review as then established can be seen as an important attempt to formulate policy by integrating science while at the same time involving public participation in the process—all within the framework of a national crisis. However, despite such extensive evidence gathering, this pathway to policy formulation is fragile, and still has a process of filtering, or refinement, in which themes emerged from the 'ground up' become overwhelmed, or sidelined, in the search for the certainty demanded of politicians for solutions.

The idea of a declaration, similar to that made by the EU in Cork in 1996, did not make the final report—not even as a minor suggestion or comment. Reasons as discussed varied, but primarily revolved around the need to have more urgent and pragmatic policy development that dealt with the drought as faced, rather than the longer-term (and perhaps more philosophical) issue of the future for rural Australia. It is the view of this author that this will prove to be a missed opportunity, one that is unlikely to re-emerge in the foreseeable future. In hindsight and given that the drought was immediately followed by two seasons of exceptional rainfall that resulted in some of the worst flooding in over five decades, the issue of the value of the countryside to the nation's future still remains an open one.[*]

Conclusions

In summary, this experience is offered as a small example of how ideas for policy responding to evidence raised in public consultation become absorbed or not on the pathway to policy development. It is also a small, but relevant, example of how the 'expert' evidence as presented by the Panel becomes lost in the overriding pressure to develop immediate and pragmatic responses, within the larger context of offering some certainty as to potential solutions, to the issue at hand. It is also salutary here to contemplate the extended time that can elapse between recommendation and action, thus challenging the assumption that the public will see the relationship between consultation and policy as proof of evidence being accepted.

At the time of writing (March 2012), for the first time in over 30 years, unusually, there is no part of Australia that is officially drought declared. For the policy makers, this means a unique opportunity to undertake policy development outside a crisis framework. It also means the end of one system of drought declaration (the exceptional circumstances process) and an historic opportunity to put an alternative system in its place. For the current minister this means that 'for future drought events the government is looking at ways of improving the current system of better help [to] *farmers* manage risks and prepare the future challenges' (Ludwig 2012, italics added).

---

[*] Again beyond this chapter but relevant is the simmering debate about the value to the nation of agricultural land as a place to grow food and fibre, or as a place to develop the underlying mineral resources (see, for example, Australian Broadcasting Corporation 2012).

On reflection, the test of the value of the evidence-gathering process will be in the future policies that become the legacy of integration of the local and expert knowledge integrated during the review consultation process. However, as the Minister's statement implies, the major challenge will be that drought, when it returns, continues to remain the responsibility of farmers and, again, is not seen as a national challenge. This means that future policy makers will again be confronted with decision making during a crisis and will perhaps once again turn to a process of national consultation to determine future action.

References

Aslin, H., and J. Russell. 2008. *Social impacts of drought: Review of the literature.* Report prepared for the Drought Review Branch of the Department of Agriculture, Fisheries and Forestry. Canberra: Commonwealth of Australia. <http://adl.brs.gov.au/data/warehouse/ClientReports/sidlrd9abps001/sidlrd9abps0010111a/SocImpctDroughtLitRev2008_1.0.0.pdf>

Australia. Productivity Commission. 2009. *Government drought support.* Report No 46. Melbourne.

Australian Broadcasting Corporation. 2012. Green group attacks Xstrata mine nod. *The World Today.* http://www.abc.net.au/news/2012-03-28/green-group-attacks-xstrata-mine-nod/3917008 (28 March 2012).

Banks, G. 2009. *Evidence-based policy making. What is it? How do we get it?* ANU Public Lecture Series Presented by ANZSOG, February 2009, Productivity Commission. Canberra: Australian Government. http://www.pc.gov.au/__data/assets/pdf_file/0003/85836/20090204-evidence-based-policy.pdf

Bishop, P., and G. Davis. 2002. Mapping public participation in policy choices. *Australian Journal of Public Administration* 61 (1): 14–29.

Bridgman, P., and G. Davis. 2000. *The Australian policy handbook,* 2nd ed. St Leonards, NSW: Allen & Unwin.

Burke, T. 2009. Government receives final drought review report. Media release by minister for agriculture, fisheries and forestry, DAFF09/213B. http://www.maff.gov.au/media/media_releases/2009/march/government_receives_final_drought_review_report (6 March 2009).

———. 2010. *Farm Household Support Amendment (Ancillary Benefits) Bill 2010: Second reading speech.* Commonwealth Parliamentary Debates, House of Representatives, 26 May 2010.

Commonwealth of Australia. 2008a. *Australia 2020 Summit overview.* Canberra.

———. 2008b. *Australia 2020. Thinking big: Future directions for rural industries and rural communities.* Canberra.

Dayton, L. 2007. Climate: The peril we face. *The Australian,* 7 April.

Drought Policy Review Task Force. 1990. *National drought policy.* Canberra: Commonwealth of Australia.

European Union. 1996. *The Cork declaration—A living countryside.* http://ec.europa. eu/agriculture/rur/cork_en.htm

Gallop, G. 2007. Certainty is a dangerous thing. *The Australian,* 24 January.

Gastil, J., and P. Levine, eds. 2005. *The deliberative democracy handbook: Strategies for effective civic engagement in the twenty-first century.* San Francisco: Jossey-Bass.

Hennessy, K., R. Fawcett, D. Kirono, F. Mpelasoka, D. Jones, J. Bathols, P. Whetton, M. Stafford Smith, M. Howden, C. Mitchell, and N. Plummer. 2008. *An assessment of the impact of climate change on the nature and frequency of exceptional climatic events.* Canberra: Bureau of Meteorology and CSIRO.

Kenny, P., S. Knight, M. Peters, D. Stehlik, B. Wakelin, S. West, and L. Young. 2008. *It's about people: Changing perspectives on dryness—A report to government by an expert social panel.* Canberra: Commonwealth of Australia.

Keogh, M., and R. Granger. 2011. *Drought Pilot Review Panel: A review of the pilot of drought reform measures in Western Australia.* Canberra: Commonwealth of Australia.

Kloppenburg, J. 1991. Social theory and the de/reconstruction of agricultural sciences: Local knowledge for an alternative agriculture. *Rural Sociology* 56 (4): 519–548.

Lawrence, C. 2010. Social sciences key to good public policy. *The Australian,* 17 November 2010.

Ludwig, J. 2012. Bundarra and Eurobodalla exceptional circumstances declarations. Media release by the minister for agriculture, fisheries and forestry, DAFF12/295L. ≤http://www.maff.gov.au/media_office/media_releases/≥ (14 March 2012).

Maiden, S. 2008. Rudd signals drought scheme overhaul. *The Australian,* 4 March.

Stehlik, D. 2009. *Considering the value of 'resilience'—Lessons from Australian droughts.* 1st Australian Rural and Remote Mental Health Symposium. Canberra, 2–3 November.

9

The Promise and Challenge of Evidence-Based Policy Making

Linda Courtenay Botterill

CONTENTS

This collection is focussed on the relationship between science and drought policy, so an important part of our discussion is inevitably covering the role of science *as evidence for* drought policy. This chapter is provided by way of reflection on the four that precede it in Section II and that specifically addressed the incorporation of science into drought policy. This discussion begins with a brief introduction to the public policy literature with its origins in the policy sciences and the goal of providing research-based advice to policy makers. This is followed by a summary of the evidence-based policy movement, its origins, and aspirations—including a critique of evidence-based policy making, in terms of what constitutes 'evidence' and also through discussion of the role that 'evidence' plays as one input into the policy process. The interaction between science, evidence, and policy is not linear but this does not mean there is not an important role for science in the development of policy. Good policy requires good evidence; however, that evidence is not always decisive in determining policy outcomes.

The Policy Sciences and Science for Policy

Social scientists have been concerned about the link between academic research and achieving good societal outcomes for over half a century. The 'policy sciences' as conceptualised by Harold Lasswell were problem oriented and explicitly value laden in that they sought to produce research for policy makers that resulted in improved outcomes for society. Lasswell (1951, 3) posed a series of questions, which remain central to the issue of evidence-based policy:

> If our policy needs are to be served, what topics of research are most worthy of pursuit?...What are the most promising methods of gathering facts and interpreting their significance for policy? How can facts and interpretations be made effective in the decision-making process itself?

The policy sciences approach implied a high degree of rationality in the policy process that persists in contemporary models of policy making, most notably in the 'policy cycle' model. However, as much as policy makers, politicians, and the general public would like to believe in the rationality and logic of the policy cycle approach, it has become clear from the research of public policy scholars and the experience of millions of public servants that, as Peter John (1998, 197) so succinctly notes, 'Reality is messy'. The first major critique of the rational aspirations of the policy sciences was the seminal article by Charles Lindblom (1959), which set out to describe how policy was actually made in the real world. He described an incremental process in which values conflicts are avoided by achieving agreement on policy means rather than policy ends and through which values and interests, overlooked in earlier iterations of the policy process, are revisited and considered in later revisiting of the policy in question. In later work, he described this process as 'serial and remedial' (Lindblom 1965, 215). More controversially, Lindblom argued that this incremental approach was also the best way that policy could be made as it recognised the limitations facing the policy maker. In a follow-up article to the original piece (1979, 159), he explained this as follows: 'The choice between synopsis and disjointed incrementalism—or between synopsis and any form of strategic analysis—is simply between ill-considered, often accidental incompleteness on one hand, and deliberate, designed incompleteness on the other'.

Lindblom was not the first to acknowledge the limits on decision makers. Herbert Simon (1953, 81) argued:

> Rationality implies a complete, and unattainable, knowledge of the exact consequences of each choice. In actuality, the human being never has more than a fragmentary knowledge of the conditions surrounding his action, nor more than a slight insight into the regularities and laws that would permit him to induce future consequences from a knowledge of present circumstances.

Simon developed the concept of 'bounded rationality', which suggests that decision makers are *intendedly* rational but only *limitedly* so' (Simon 1957, xxiv). Jones (1999, 299) explains this intentionally limited rationality as follows: 'Like comprehensive rationality, bounded rationality assumes that actors are goal-oriented, but bounded rationality takes into account the cognitive limitations of decision makers in attempting to achieve those goals'.

Since Lindblom's 1959 article appeared, there has been ongoing debate in the academic literature about the possibility of rational approaches to policy (for example, Arrow 1964; Dror 1964; Etzioni 1967; Smith and May 1993; Bridgman and Davis 2003; Howard 2005). Although scholars generally agree that incrementalism is an accurate description of the real world of the policy maker, they have not necessarily agreed with Lindblom's advocacy of incrementalism as a normative model. Over the past half century, other models of the policy process have been developed that recognise the essential messiness of policy making, the role of the various players in the policy process (Richardson and Jordan 1979; Sabatier 1988), the interplay of values (Thacher and Rein 2004; Stewart 2006), the role of institutions (Thelen 2003; Pierson 2004; Streeck and Thelen 2005), and the challenges of tackling 'wicked problems' (Rittel and Webber 1973; APSC 2007; Head 2008).

While most public policy scholars today accept the limitations on the rational policy process described by Simon, Lindblom, and others, there remains reluctance to give up completely the aspiration of rational policy making. The most recent manifestation of the desire to improve the policy process and to integrate science, understood broadly, into policy is the evidence-based policy movement. The emergence of evidence-based policy making as an aspiration in the past 15 years or so has been seen by some scholars (for example, Parsons 2002; Williams 2002; Boaz et al. 2008), including the present author (Botterill and Hindmoor 2012), as a misguided attempt to revive the rationalist dream and to rescue policy making from the pathologies of value-laden incrementalism through the incorporation of evidence into the policy process. Parsons (2002, 45) colourfully describes the evidence-based policy movement as 'not so much a step forward as a step backwards; a return to the quest for a positivist yellow brick road—somewhere, over Charles Lindblom'.

The Aspirations and Realities of Evidence-Based Policy Making

The evidence-based policy movement has its intellectual origins in evidence-based medicine and its major political genesis under the Blair Labour government in the United Kingdom in the late 1990s. The Labour Party sought to move away from what was seen as an ideological determination of policy toward a focus on 'what works' (Clarence 2002, 2). This was seen as pragmatic policy making, freed from value-laden politics. The approach was twofold: using pilot studies of policy approaches and evaluating them for evidence of their efficacy, and drawing on research to provide guidance on effective policy. Importantly, the approach included social scientific evidence.

However, critics identify some shortcomings in the concept. First, they are concerned that particular types of evidence are privileged over other sources of information. Although social science is considered to be important, preference appears to be given to research that mimics the physical sciences, such as the use of randomised controlled trials as against more qualitative forms of research. As Stehlik points out in Chapter 8, evidence collected through consultation processes is often considered even less authoritative. It should be noted, though, that consultation is an important part of the policy process in Australia, with bodies like the economically focussed Productivity Commission including calls for public submissions and the holding of public hearings as regular features of their inquiry processes. Similarly, Knutson and Haigh note in Chapter 11 the 'listening sessions' held by the US National Drought Policy Commission.

Second, the evidence-based policy approach overlooks the fact that, in spite of attempts by policy makers to 'remove the politics from policy', policy making is a political activity and as such is inherently about values: the values of policy makers and the values of different groups in society interested in a particular policy issue. These values do not always align, so an important role for government is determining trade-offs between conflicting values positions (Lindblom 1965, 227). Various strategies are employed by policy makers in achieving this balancing act (Stewart 2006; Thacher and Rein 2004) but the outcome is inevitably a form of values compromise. Research-based evidence will be factored into the policy process but it will not necessarily be decisive. Some policy issues involve deeply held personal and community values and, in these cases, scientific evidence is less likely to be influential (see Pielke 2007 for a clear discussion of the limited role scientific advice can play in values-based debates).

Apart from the policy process itself being essentially about values, it should be noted that the evidence put forward by researchers is not value free. As has been argued elsewhere (Botterill and Hindmoor 2012), the inevitable simplification of science that takes place in the transmission of scientific research from the journals to policy debate involves value judgements. Just as operating within an environment of bounded rationality is the reality for policy makers, bounded rationality is also present in the transmission of evidence. As Fleck (1979 [1935], 115) explains, some sort of simplification process is unavoidable:

> Every communication and, indeed, all nomenclature tends to make any item of knowledge more exoteric and popular. Otherwise each word would require a footnote to assign limitations and provide explanations. Each word of the footnote would need in turn a second word pyramid. If continued, this would produce a structure that could be presented only in multidimensional space. Such exhaustive expert knowledge completely lacks clarity and is unsuitable.

The choices that are made about which bits of information matter and which do not involve values-based judgements. In addition to the implications of the simplification process, evidence is not always complete or uncontested, resulting in a further exercise of judgement in the face of uncertainty and the management of risk. Policy makers cannot possibly have access to or absorb every piece of information or evidence that is relevant to the policy problem at hand.

The evidence-based policy making (EBPM) push has raised some important questions about what constitutes 'evidence'. Parsons (2002, 46) argues that 'in EBPM what is to count is what can be counted, measured, managed, codified and systematised'. The literature also points to a preference for the systematisation of science (see, for example, Pawson 2006; Bulmer et al. 2007; Boaz et al. 2008) to provide the scientific consensus on a particular issue in accessible and digestible form for policy makers. This partly reflects the origins of EBPM in evidence-based medicine but also reveals a bias toward positivist approaches to research. Pawson (2006, 15–16) notes:

> What we have at present is:
>
> 1 A rather narrow experimental trial-based model of valid knowledge and a statistical model of accumulation, which risk ignoring vital evidence that might be gained from other means and sources.
> 2 An ambition, perhaps no more than wishful thinking, about synthesizing a plurality of data, which risks glossing over the considerable epistemological divides on which different knowledge perspectives are founded.

Nutley et al. (2009, 20) suggest that there needs to be broader conception of what counts as 'evidence'. The type of interactive policy making that they suggest in place of the instrumental, linear model of evidence-based policy making brings the debate about the policy process full circle; policy making is values based, contested, messy, and ultimately a process of negotiation. As Head (2010, 83) has pointed out

> Policy decisions in the real world are not deduced from empirical-analytical models, but from politics and practical judgement. There is an interplay of facts, norms and preferred courses of action. In the real world of policy-making, what counts as 'evidence' is diverse and contestable. The policy-making process in democratic countries uses the rhetoric of rational problem-solving and managerial effectiveness, but the policy process itself is fuzzy, political and conflictual.

Australia's experience with the implementation of its national drought policy illustrates the essential messiness, and political nature, of the policy process. Designed to end ad hoc response to drought events, the policy has not been able to escape the 'politics and practical judgement' to which Head refers, in spite of attempts to bring science into the process of declaring exceptional drought and thereby triggering government assistance. Likewise, the experience in the United States of the development of increasingly sophisticated monitoring tools has failed to be translated into a national policy approach based on risk management and mitigation. It is not clear that these tools have been taken up effectively by their intended end users and the risk management intention of these tools has to some extent been undermined by the tendency of the US government to use the Drought Monitor as a trigger for a plethora of reactive assistance programmes.

Key members of the drought science community in the United States have taken on the role of 'policy entrepreneur' (Kingdon 1995; Botterill forthcoming) but, in the absence of a political champion, have not succeeded in achieving large-scale, national policy change. The limited success encountered by these policy entrepreneurs is illustrative of a key feature of Kingdon's model of the policy process: the need to get the timing right and to exploit a window of opportunity when the political and policy streams come together. For the scientist in the United States, these windows of opportunity appeared to open with various bills before the Congress, only to be overtaken by other events in the politics stream, such as the 9/11 terrorist attacks. Kingdon's description of the policy world emphasises the short-lived nature of these windows of opportunity (Kingdon 1995, 160) and the importance of available solutions being on hand when particular problems become the focus of policy attention.

The Way Forward

The preceding discussion should not be interpreted as an argument for leaving policy determination entirely to the forces of power and influence without reference to research-based evidence. Good quality evidence clearly has an important role in informing policy, in clarifying the trade-offs that are being made, in providing reliable information to those engaged in policy debate, and in potentially raising the quality of public policy deliberation. Young et al. (2002, 223) argue that 'not evidence-based *policy*, but a broader *evidence-informed society* is the appropriate aim'. In this form, research should be focussed less on solving policy problems and more on 'clarifying issues and informing the wider public debate' (Young et al. 2002, 218). This focus on research informing the wider policy debate beyond simply providing an input into a linear policy process is supported by other authors (for example, Williams 2002). The bottom-up approach to drought information taken in the United States through NIDIS and the accessibility of reliable and up-to-date drought information through the Drought Monitor are consistent with this evidence-informed society approach.

Australia's top-down approach to drought policy making has not demonstrated these characteristics. The consultation process described by Daniela Stehlik in the preceding chapter was focussed on drought-affected communities as the objects of research, not as its audience. While the inclusion of social research in the 2008 drought review was innovative, it still demonstrated the characteristics of a linear process in which the communities were studied and the outcomes of the study were fed into the policy process. The short-lived National Agricultural Monitoring System was closer to a model of providing evidence *to* the community but it remained constrained in its objectives. It was focussed on providing access to information by communities to assist them in the narrow task of making a case that government support was warranted, within the confines of existing policy settings. It was not primarily about providing evidence to increase the community's knowledge base more generally to generate better informed drought policy debate.

Evidence clearly has a role to play in the development of drought policy both in Australia and the United States, and diversity of evidence will contribute to better policy. The inclusion in the Australian government's drought policy review in 2008 of an expert social panel redressed a serious oversight in earlier policy making that had largely focussed on the economics of farm businesses. Similarly, including a review of the likely impact of climate change on future drought severity and frequency (Hennessy et al. 2008) was important in informing policy makers about future climatic challenges facing agricultural producers. The experience of our authors, writing as academics who have also operated as insiders in drought policy development, highlights some of the potential, challenges, and limitations of an evidence-based approach to policy making.

References

APSC. 2007. *Tackling wicked problems: A public policy perspective.* Canberra: Commonwealth of Australia. http://www.apsc.gov.au/publications07/wickedproblems.htm

Arrow, K. J. 1964. Review: A strategy of decision: Policy evaluation as a social process by D. Braybrooke and C. E. Lindblom. *Political Science Quarterly* 79 (4): 584–588.

Boaz, A., L. Grayson, R. Levitt, and W. Solesbury. 2008. Does evidence-based policy work? Learning from the UK experience. *Evidence and Policy* 4 (2): 233–253.

Botterill, L. C. Forthcoming. Are policy entrepreneurs really decisive in achieving policy change? The development of drought policy in the USA and Australia. *Australian Journal of Politics and History.*

Botterill, L. C., and A. Hindmoor. 2012. Turtles all the way down: Bounded rationality in an evidence-based age. *Policy Studies.* doi: 10.1080/014-42872.2011.626315.

Bridgman, P., and G. Davis. 2003. What use is a policy cycle? Plenty, if the aim is clear. *Australian Journal of Public Administration* 62 (3): 98–102.

Bulmer, M., E. Coates, and L. Dominian. 2007. Evidence-based policy making. In *Making policy in theory and practice*, ed. H. Bochel and S. Duncan, 87–104. Bristol: The Policy Press.

Clarence, E. 2002. Technocracy reinvented: The new evidence based policy movement. *Public Policy and Administration* 17 (1): 1–11.

Dror, Y. 1964. Government decision making muddling through—'Science' or inertia? *Public Administration Review* XXIV (3): 153–157.

Etzioni, A. 1967. Mixed-scanning: A "third" approach to decision-making. *Public Administration Review* XXVII (5): 385–392.

Fleck, L. 1979 [1935]. *Genesis and development of a scientific fact*. Translated by T. J. Trenn and R. K. Merton. Chicago: University of Chicago Press.

Head, B. 2008. Wicked problems in public policy. *Public Policy* 3 (2): 101–118.

———. 2010. Reconsidering evidence-based policy: Key issues and challenges. *Policy and Society* 29:77–94.

Hennessy, K., R. Fawcett, D. Kirono, F. Mpelasoka, D. Jones, J. Bathols, P. Whetton, M. Stafford Smith, M. Howden, C. Mitchell, and N. Plummer. 2008. *An assessment of the impact of climate change on the nature and frequency of exceptional climatic events*. Canberra: Bureau of Meteorology and CSIRO.

Howard, C. 2005. The policy cycle: A model of post-Machiavellian policy making? *Australian Journal of Public Administration* 64 (3): 3–13.

John, P. 1998. *Analysing public policy*. London: Pinter.

Jones, B. D. 1999. Bounded rationality. *Annual Review of Political Science* 2:297–321.

Kingdon, J. 1995. *Agendas, alternatives and public policies*, 2nd ed. New York: Longman.

Lasswell, H. D. 1951. The policy orientation. In *The policy sciences*, ed. D. Lerner and H. D. Lasswell, 3–15. Stanford: Stanford University Press.

Lindblom, C. E. 1959. The science of 'muddling through'. *Public Administration Review* 19:79–88.

———. 1965. *The intelligence of democracy: Decision making through mutual adjustment*. New York: Free Press.

———. 1979. Still muddling, not yet through. *Public Administration Review* 39 (6): 517–525.

Nutley, S., I. Walter, and H. Davies. 2009. Past, present and possible futures for evidence-based policy. In *Evidence for policy and decision-making: A practical guide*, ed. G. Argyrous, 1–23. Sydney: UNSW Press.

Parsons, W. 2002. From muddling through to muddling up—Evidence-based policy making and the modernisation of British government. *Public Policy and Administration* 17:43–60.

Pawson, R. 2006. *Evidence-based policy: A realist perspective*. London: Sage Publications.

Pielke, R. A., Jr. 2007. *The honest broker: Making sense of science in policy and politics*. Cambridge: Cambridge University Press.

Pierson, P. 2004. *Politics in time: History, institutions, and social analysis*. Princeton, NJ: Princeton University Press.

Richardson, J. J., and A. G. Jordan. 1979. *Governing under pressure: The policy process in a post-parliamentary democracy*. Oxford: Robertson.

Rittel, H. W. J., and M. M. Webber. 1973. Dilemmas in a general theory of planning. *Policy Sciences* 4:155–169.

Sabatier, P. 1988. An advocacy coalition framework of policy change and the role of policy-oriented learning therein. *Policy Sciences* 21:129–168.

Simon, H. A. 1953. *Administrative behaviour: A study of decision-making processes in administrative organisation.* New York: The Macmillan Company.

———. 1957. *Administrative behaviour: A study of decision-making processes in administrative organisation.* New York: The Macmillan Company.

Smith, G., and D. May. 1993. The artificial debate between rationalist and incrementalist models of decision making. In *The policy process: A reader,* ed. M. J. Hill, 197–211. New York: Harvester-Wheatsheaf.

Stewart, J. 2006. Value conflict and policy change. *Review of Policy Research* 23 (1): 183–195.

Streeck, W., and K. Thelen, eds. 2005. *Beyond continuity: Institutional change in advanced political economies.* Oxford: Oxford University Press.

Thacher, D., and M. Rein. 2004. Managing value conflict in public policy. *Governance* 17 (4): 457–486.

Thelen, K. 2003. How institutions evolve: Insights from comparative historical analysis. In *Comparative historical analysis in the social sciences,* ed. J. Mahoney and D. Rueschemeyer, 208–240. Cambridge: Cambridge University Press.

Williams, N. 2002. Evidence and policy: Towards a new politico-administrative settlement. *Political Quarterly* 73 (1): 86–97.

Young, K., D. Ashby, A. Boaz, and L. Grayson. 2002. Social science and the evidence-based policy movement. *Social Policy and Society* 1 (3): 215–224.

Section III

From Policy to Action

10

Developing Early Warning and Drought Risk Reduction Strategies

Chad A. McNutt, Michael J. Hayes, Lisa S. Darby,
James P. Verdin, and Roger S. Pulwarty

CONTENTS

Drought is one of the least understood of the natural hazards and also one of the costliest (Wilhite 2000; National Climatic Data Center [NCDC] 2011). Having an early indication that drought will develop or intensify is critical to employing timely strategies that can mitigate and reduce its impacts. In its most basic form, drought can be thought of as insufficient water to meet demand. Demand can be based on ecosystem processes or on institutional and economic systems linked to human health and welfare. Because understanding demand is critical, systems designed to provide early warning of drought ideally should be able to evaluate changes in both the demand for, and supply of, water and successfully communicate the information to groups or institutions that can apply drought risk reduction strategies.

Perhaps there is no more contentious common resource than water. Humorists have observed that water can defy the laws of physics (that is, by flowing uphill toward money); in some jurisdictions, it has been the subject of intense and long-term litigation and its extremes have been responsible for devastating disasters and subsequent population migration and resettlement. How does early warning operate, then, in a potentially contentious environment where 'good' information may not be enough to overcome political, institutional, and other barriers to action? Numerous theoretical and empirical studies have shown that effective management of common-pool resources is easier to realise when communities develop social capital through a distributed and dense social network that develops trust and common understanding and stimulates learning and formulation of alternative response options (see, for example, Jodha 1990; Bray et al. 2003; Pretty and Smith 2004; Adger et al. 2005; Cash et al. 2006).

Managing contentious issues of this nature is particularly challenging in federal-state systems and these challenges can be considered at three levels:

- Constitutional arrangements and divided responsibilities can hamper decision making around shared resources.
- Institutions within jurisdictions can engage in 'turf wars', 'buck passing', or other forms of bureaucratic politics.
- Effective planning requires engagement of affected communities in decision making.

This chapter provides two case studies of efforts by the National Integrated Drought Information System (NIDIS) and the National Drought Mitigation Center (NDMC) to work consultatively with different levels of government and community groups to enhance drought early warning systems. The chapter provides an outline of the history and complementary nature of the two entities and describes their collaborative partnership through research-based knowledge assessments and prototyping of regional early warning information systems in support of climate risk management. It then focuses on two case studies that highlight the political and institutional challenges facing the development of these systems. The chapter begins with an overview of the state of play of drought planning in the United States.

Drought Planning in the United States

Within the United States, drought planning is taking place at a variety of institutional levels. During the past three decades, in the absence of any noteworthy federal-level drought planning, significant progress has been made by individual states. Figure 10.1 shows the status of drought planning at the state level within the United States as of December 2011. In 1982, three states had drought plans (Colorado, South Dakota, and New York); by December 2011 this had increased to forty-six states having some kind of state-level drought plan or one in development, though the specific focus of the plans varies between states.

In 1995, Montana took the important step of incorporating drought mitigation actions into its drought plan. New Mexico received funding from

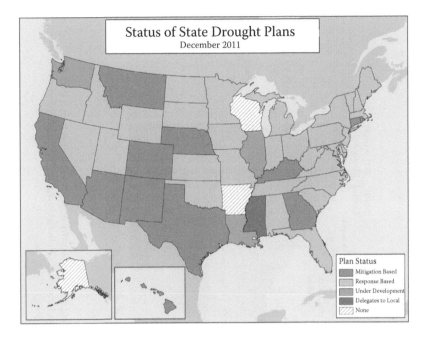

FIGURE 10.1
The status of state-level drought planning within the United States as of December 2011.

the US Bureau of Reclamation and, after working directly with the National Drought Mitigation Center, completed its state-wide drought mitigation plan in 1998. Since then, ten states have followed Montana's example and have either revised their existing plans to incorporate mitigation or developed new plans that include mitigation actions. Figure 10.2 illustrates how the progress of state-level planning across the country has benefited from the windows of opportunity resulting from focusing events such as droughts or the NDMC's formation.

Drought mitigation planning has also taken place in other venues within the United States, with Native American nations taking a strong interest in developing mitigation plans. In the south-western United States, the Navajo, Hualapai, Hopi, and Zuni tribes, as well as the Taos Pueblo, have all recently drawn up drought mitigation plans.

National Drought Mitigation Center (NDMC)

The NDMC was established in 1995 at the University of Nebraska-Lincoln. Its mission is to help people and institutions develop and implement measures to reduce societal vulnerability to drought through risk management.

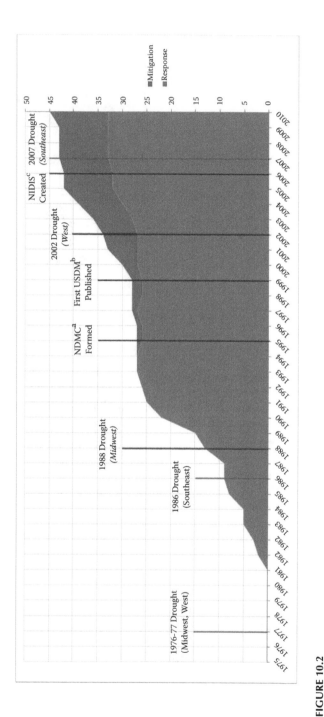

FIGURE 10.2

Time line associated with the number of drought plans within the United States, along with noteworthy droughts and events that may have had an impact on the progress of planning. [a] National Drought Mitigation Center. [b] US Drought Monitor. [c] National Integrated Drought Information System.

The NDMC has worked extensively on a variety of drought risk management topics with local, state, regional, federal, and tribal governments and agencies in the United States, as well as several international organisations. Because of the diverse and complex nature of drought, the NDMC has employed an interdisciplinary group of faculty and staff over the years with backgrounds in climatology, meteorology, hydrology, water resources, geography, remote sensing and geographic information systems, economics, anthropology, rural sociology, public participation, and journalism in order to accomplish its mission.

The NDMC plays a prominent role in several national monitoring efforts, including the US and North American Drought Monitor products. The current state-of-the-art national drought monitoring tool for the United States is the US Drought Monitor (USDM), which integrates traditional hydrological and climate-based drought indices, remote sensing-based vegetation indices, in situ observations (for example, stream flow and soil moisture), and input from local experts (Svoboda et al. 2002). The USDM is a national-level map generated weekly that reflects the magnitude, spatial extent, and impact of drought. The USDM represents a collective effort between the NDMC, National Oceanic and Atmospheric Administration (NOAA), and US Department of Agriculture (USDA) to develop a single, meaningful drought classification process for the country.

The US Drought Monitor incorporates feedback and input into the process by maintaining and utilising an expert user group of approximately 300 people from across the United States who provide ground truth for the indicators. A convergence of evidence approach is used to combine the indices with impacts and feedback from experts through an iterative process each week. The classification system is based on the utilisation of a ranking percentile approach. This approach gives historical context to any index value in that it shows the percentage of values in its frequency distribution and thus allows comparison of multiple indices. The classification categories run from D0 to D4—where D0 is equal to 'abnormally dry' (thirtieth percentile) conditions and is not a drought category but signifies the potential for drought, D1 is considered 'moderate drought' (twentieth percentile), D2 is 'severe drought' (tenth percentile), D3 is 'extreme drought' (fifth percentile), and D4 is considered 'exceptional drought' (second percentile).

Since its inception in 1999, the USDM has gained widespread acceptance as a primary source of drought information in the United States and has been used by federal- and state-level programmes (for example, USDA Farm Service Agency's Livestock Forage Disaster Program, state drought plan triggers, and the Internal Revenue Service) and various media outlets (including the Weather Channel; newspapers such as the *New York Times, Wall Street Journal,* and *USA Today;* network news channels; CNN; and various web-based print and radio sources including National Public Radio). The USDM continually evolves as new tools and data sets become available in an effort to support the growing demands of a diverse user community for drought information. The USDM was originally intended to provide a broad-scale snapshot of drought conditions across the United States, but there is growing interest for the USDM to provide county (local government)-level drought information to accommodate a wide range of local-scale decision-making activities.

The NDMC has worked closely with several federal agencies from its beginning, often serving as a liaison between federal and state authorities. The linkages between the NDMC and the National Oceanic and Atmospheric Administration (NOAA) have been strong; most recently, the NDMC has been interacting with the regional climate service directors in NOAA's eastern, central, and southern regions on activities related to recent drought events.

National Integrated Drought Information System (NIDIS)

One primary connection between the NDMC and NOAA has been with the National Integrated Drought Information System (NIDIS). The NDMC was a key participant in the conceptual development of NIDIS, and has since been heavily involved in a variety of NIDIS activities and leadership opportunities. In 2004, following discussions involving the establishment of a national drought policy, the Western Governors' Association (WGA)* proposed an idea for the creation of NIDIS. These discussions followed on the heels of the driest 5-year period in the Colorado Basin over the period of record. The NIDIS Act of 2006 (Public Law 109-430) prescribes an interagency approach, led by NOAA, to 'enable the nation to move from a reactive to a more proactive approach to managing drought risks and impacts'. NIDIS is authorised to

- provide an effective drought early warning system that both collects and integrates information on the key indicators of drought and drought severity, and provides information that reflects state and regional differences in drought conditions
- coordinate federal research in support of a drought early warning system
- build upon existing forecasting and assessment programs and partnerships

---

* WGA is made up of the governors of nineteen states and three US-flag Pacific islands: namely, Alaska, American Samoa, Arizona, California, Colorado, Guam, Hawaii, Idaho, Kansas, Montana, Nebraska, Nevada, New Mexico, North Dakota, Northern Mariana Islands, Oklahoma, Oregon, South Dakota, Texas, Utah, Washington, and Wyoming.

The governance arrangements for NIDIS include representatives from federal, state, and Native American tribal agencies, and academic and private entities.

NIDIS supports or conducts impacts assessment, forecast improvements, indicators and management triggers, and the development of watershed-scale (web-based) information portals. In partnership with other agencies, tribes, and states, NIDIS coordinates and develops capacity to prototype and then implement regional drought early warning information systems using the information portals and other sources of local drought knowledge. NIDIS also conducts knowledge assessments to determine where major gaps in data, forecasts, communication, and information delivery exist; to identify innovations in drought risk assessment and management at state and local levels; and to engage constituents in improving the effectiveness of NIDIS. In recent years, these knowledge assessments have covered the status of drought early warning systems in the United States, satellite remote sensing of drought, drought indices and hydrologic data web services, soil moisture observation networks, and tribal early warning and drought preparedness.

An important aspect of NIDIS's work is the facilitation of communication and coordination. This is achieved through mechanisms such as 'webinars', workshops, and the Engaging Preparedness Communities Working Group. The success of NIDIS depends overall on the significant leveraging of existing system infrastructure, data, and products produced by operational agencies (for example, the Natural Resources Conservation Service snow-depth network and reservoir levels from federal agencies such as the Department of the Interior and the US Army Corps of Engineers).

Building a US Drought Early Warning Information System

As noted before, the development of early warning systems for drought is critical to effective drought mitigation and risk management. NIDIS is developing a network of regional drought early warning systems (RDEWS) by building on existing monitoring and forecast products and service networks, such as the US Drought Monitor (USDM) and the National Weather Service's 90-day seasonal outlook. NIDIS increases the value of such existing products through the provision of improved communication and coordination of monitoring, forecasting, and impact assessment efforts at national, river basin, state, and local levels. The ultimate objective of NIDIS is to provide a better understanding of how and why droughts affect society, the economy, and the environment, and to improve accessibility, dissemination, and use of early warning information for drought risk management. To build the regional drought early warning information systems, NIDIS is undertaking several pilot projects to prototype and develop a platform to support the communication among climate service providers (for example, regional climate centres, State Climatologist, National Climatic Data Center), scientists, community members, and policy makers for testing knowledge management and use across the full spectrum of climate timescales (that is, seasonal, interannual, decadal, and centennial scales). Combining different forms of knowledge, whether they are informal, theoretical, or experimental, can enhance legitimacy, integrity, and the overall value of the information. The following case studies describe the work of NIDIS in the Upper Colorado River Basin and the Apalachicola–Chattahoochee–Flint River Basin. The analysis centres on how networks and partnerships in these two US river basins have evolved around drought monitoring through the US Drought Monitor and NIDIS and the potential impact on improved communication and transfer of information, building of trust and legitimacy, and overall community engagement that translates into drought early warning and enhanced risk reduction strategies.

NIDIS Upper Colorado River Basin Pilot

The Colorado River Basin (CRB) drains an area of 637,000 square kilometres that includes parts of seven western US states (Wyoming, Colorado, Utah, New Mexico, Nevada, Arizona, California) and Mexico (Figure 10.3). Three-quarters of the Colorado basin comprises national forests, national parks, and Indian reservations. Approximately two-thirds of the water flowing in the Colorado River and its tributaries are used for irrigation, while the other third supplies urban areas, sustains ecosystems, or is lost through evaporation. The CRB is a major source of water for the southwest and its urban centres. Throughout the twentieth century several dams were built on the Colorado and its tributaries. The primary purposes of the dams are water supply, electricity generation, flood control, and recreation and tourism. The basin dams can store over 86 billion cubic metres of water—approximately four times the Colorado River's average annual runoff. The largest project in the basin is Hoover Dam, located on the border between Nevada and Arizona. The second largest dam is Glen Canyon Dam, in northern Arizona. The two dams combined provide approximately 80 percent of the water-storage capacity in the basin.

In 1922 the Colorado River Compact was signed, which has become a milestone in the management of the river and the foundation for the apportionment of the river for the seven Colorado River Basin states. The river was administratively divided at Lees Ferry, Arizona, into the lower basin states—Arizona, Nevada, and California—and the upper basin states—Wyoming, Utah, Colorado, and New Mexico. The total annual flow of the Colorado River was estimated at the time to be approximately 21,000 million cubic metres at Lees Ferry, of which 18,500 million cubic metres were divided between the lower and the upper compact states. It was later discovered, however, that the initial estimate of Colorado River annual runoff was overly optimistic

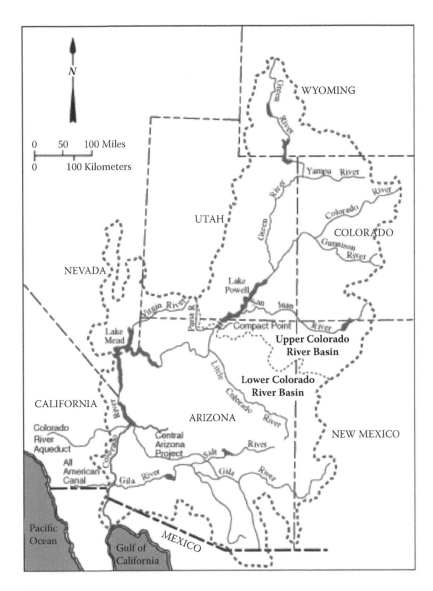

FIGURE 10.3
Map of Colorado River Basin. (Bureau of Reclamation: http://www.usbr.gov/lc/region/g4000/contracts/watersource.html.)

and based on an abnormally wet period. The actual long-term average is closer to 18,000 million cubic metres (Woodhouse, Gray, and Meko 2006).

Not surprisingly, the distribution of river flows has been the cause of interstate conflict and litigation. The impact of dams and diversions in the Colorado River Basin has generated disagreement over river development and the maintenance of ecological in-stream flows. Given projected growth in the region, these controversies and debates will continue into the future and will pose difficult trade-offs for water managers, politicians, and their constituents. One of the areas where conflict has been observed most recently is the sale, lease, and transfer of agricultural water rights to municipalities, particularly in Southern California and Colorado. Irrigated crop production accounts for approximately 80 percent of western US water supplies, a resource that some perceive as one of the most important, and potentially last, large supplies that could be converted for urban use.

In spite of these political challenges, however, over the last decade there have been positive collaborative steps taken among the CRB states. In 2002 the region underwent a severe drought, which led many to consider more seriously the real potential for future water shortages. As a result of the apparent increase in drought frequency and severity in the 1990s and 2000s, the Colorado River Basin states have increased their level of cooperation. In December 2007, the Secretary of the Interior approved guidelines for reservoir operations during drought that minimise shortages in the lower basin and risk of curtailment in the upper basin, while encouraging conservation. The basin states had substantial input into the environmental impact statement process that governed guideline development, along with tribes and environmental NGOs. The guidelines have been widely hailed as a major step forward in the CRB.

While the interstate coordination was a positive first step at coping with future droughts and growing water demands, increased communication and collaboration between the scientific and water management communities is equally important. A report by the National Academy of Sciences' National Research Council (2007) suggested that the research and information available on the CRB could be utilised more effectively through a better commitment to communication between scientists and water managers, leading to improved awareness, preparedness, and planning for drought and other water shortages. It is in this context that the NIDIS upper Colorado River Basin (UCRB) pilot was conceived through which NIDIS embarked on a process to promote collaboration among a large group of information users and providers in the UCRB. This was initiated through a series of workshops, interviews, and focus groups that involved federal, state, and local agencies; universities; and NGOs. A fundamental premise of NIDIS is that enhanced coordination and access to information will lead to improved planning and management of drought impacts, compared to previous events in the region. Through the workshops, several priorities were identified that NIDIS should consider in the pilot. They included an assessment of gaps in present monitoring and forecasting systems within the UCRB; assimilation of existing drought-related indicators, triggers, and trends into one accessible location; improved interaction (existing websites, data sets) with the US Drought Portal* to begin developing a Colorado basin drought portal and

_____
* www.drought.gov

information clearinghouse; and the investigation of the development of an upper Colorado basin-specific drought monitor (including consideration of interbasin transfers and ecosystem impacts).

One of the first activities undertaken was to engage the Colorado state climatologist to conduct a survey of the indicators and triggers currently being used in the upper Colorado River Basin. State climatologists are usually appointed by state governments and serve through a state university. In the state of Colorado, the state climatologist has strong connections with state drought planners in addition to local decision makers such as water conservancy districts. The triggers and indicators survey was necessary to understand how people in the basin perceived their vulnerability and risks related to drought, and what evidence leads them to take action or change their management activities. What the survey found was that most decision makers use drought indices but not necessarily to trigger decisions. Instead, it was more likely an index was cited as evidence to a governing authority or board to justify why a given decision was made. The survey also found that the US Drought Monitor was not seen as locally relevant; however, most people surveyed liked the simple snapshot approach and used it to keep updated on drought in surrounding areas of Colorado or the region. The indicators and triggers survey was successful at showing which indicators people used in the upper Colorado River Basin and how they are actually used to make decisions. Based on the information in the survey and the previous stakeholder meeting, evidence was building for the development of a drought monitor specific to the UCRB that would incorporate indicators and indices that are meaningful to the basin since most people surveyed believed the US Drought Monitor did not accurately capture local conditions.

Following the indicators and triggers survey, the Colorado state climatologist, who leads the Colorado Climate Center (CCC) at Colorado State University, started weekly drought assessment webinars. The purpose of the webinars was to bring people together from the basin to discuss conditions on an ongoing basis to improve awareness. The webinars started in early 2010 and brought together people from federal and state agencies, water conservation districts, recreation and tourism, and others to discuss status of the snowpack, stream flow, reservoir conditions, water demand (based mostly on temperature), and both short-term (5 days) and seasonal (30–90 days) forecasts.

From the beginning it was recognised that the upper Colorado River Basin webinars would benefit from having the US Drought Monitor author of the week participate and hear the discussion of local conditions or impacts and that this would inform how drought is depicted over the upper Colorado River Basin and Colorado. One of the direct benefits of this CCC–USDM interaction is the improved level of awareness that is generated out of the weekly drought assessment webinars. The fact that drought conditions are continuously monitored has had downstream benefits to the Colorado Water Availability Task Force (WATF), which is charged with monitoring conditions that could lead to water supply issues in Colorado. If the WATF determines drought conditions have reached a specific level, then the WATF can work with the Drought Task Force (chaired by the directors of the Departments of Agriculture, Natural Resources and Local Affairs) to make a recommendation to the governor of Colorado to activate the Colorado Drought Mitigation and Response Plan and any of the six impact task forces that determine both the vulnerability and impacts of a drought to sectors such as agriculture, municipal water, tourism, wildlife, and wildfire. The Colorado state climatologist participates in the WATF and has used the weekly webinars to provide early warning of potential impacts to the impact task forces leads when certain impact indicators start to develop.

Outside Colorado there are benefits to having a national product or map, such as the USDM, reflect local conditions. This is because the USDM designations are used as a trigger by several programs in the US Department of Agriculture (USDA) for agriculture assistance programs and by the Internal Revenue Service for tax deferrals for livestock producers that involuntarily sold livestock due to drought conditions. By providing local input into the USDM, livestock producers in particular can have more trust in the accuracy of the product despite the fact that better local information may not always actually translate into more assistance funding (for a discussion of the tendency for indicators to be used as mechanisms for triggering applications for assistance and/or government support, see Botterill and Hayes 2012).

Considering the obvious benefits of the USDM guiding federal assistance programs on the one hand and informing state drought planning efforts on the other, a picture begins to emerge of partnerships informing the USDM process from multiple levels and that the mixture of information and incentives could be considered mutually reinforcing. In other words, as the US Drought Monitor improves its depiction of drought by incorporating locally generated data, it also improves credibility and legitimacy at the federal level with USDA and the Internal Revenue Service as an authoritative source for depicting drought. At the same time, credibility and legitimacy are improved at the state level because it is being considered for the state's drought plan and because actors in the state are actively participating in the USDM process. Direct benefits to the states are important as they will likely be critical to sustaining the local networks of both people and monitoring systems, such as state mesonets* that will provide state level monitoring input to the US Drought Monitor.

Integrating networks and forming partnerships in the upper Colorado River Basin at least has initially led to better communication among stakeholders. The benefits have potentially improved state drought early warning and improved communication to its impact task forces. Through the

---

* Mesonets are typically a series of automated weather stations designed to observe mesoscale meteorological phenomena. The automated weather stations are usually operated and maintained by state agencies and/or state universities.

interaction between the USDM and the Colorado Climate Center, mutual rewards and benefits for building a cooperative foundation were observed and could serve to sustain the interaction in the long term.

What has been less successful is integrating groups outside Colorado to the broader UCRB community. The UCRB incorporates the states of Wyoming and Utah and to date the level of activity observed in Colorado has not been replicated in the other two states. It will be interesting to see if this continues, especially as the success of Colorado's interaction is observed across the basin. Another possibility for the limited engagement with Wyoming and Utah is that, while it is intuitive to work on the level of a basin or watershed, the governance structures do not always reinforce this approach. More work will be needed to engage both states and explore ways the interaction could improve overall drought risk management in the UCRB.

NIDIS Apalachicola–Chattahoochee–Flint River Basin Pilot

The Apalachicola–Chattahoochee–Flint (ACF) River Basin encompasses parts of three US states: Alabama, Florida, and Georgia (Figure 10.4). The ACF River Basin provides many benefits and services to the region's residents, municipalities, farms, and other economic sectors, and ecosystems. The Chattahoochee River originates in the Blue Ridge Mountains of the Appalachian Highlands of northeast Georgia, where it flows south-southwest along the Georgia and Alabama borders. The Flint River originates south of Atlanta in the Piedmont Province and flows southerly to the upper Coastal Plain, where it joins the Chattahoochee River in Lake Seminole to form the Apalachicola River, which flows into Apalachicola Bay and the Gulf of Mexico. The ACF has a number of multiple purpose reservoirs constructed by the US Army Corps of Engineers and by nonfederal entities for flood control, water supply, power, and commercial navigation. The majority of the dams in the ACF River Basin are on the Chattahoochee River. Dam construction in the basin began in the early 1800s on the Chattahoochee River above the fall line at Columbus, Georgia, to take advantage of natural elevation gradients for power production. During low flow periods, stored water is used to supplement the discharge of the river. Buford Dam, which impounds Lake Lanier, is the largest reservoir in the ACF system and provides 65 percent of the water storage to regulate flows; yet, it only drains 5 percent of the ACF River Basin. There have been marked decreases in the frequency of high and low flows since the start of operation of Buford Dam in 1956. Given the topography of the region, the upper reaches of the basin can be thought of in terms of hydropower and water supply, whereas the lower reaches, particularly below the fall line where the piedmont meets the coastal plain, feature several threatened and endangered species and agriculture (corn, cotton, and peanut), which relies heavily on irrigation that is supplied largely by the Floridan Aquifer.

The politics of the basin's management are typical of the problems faced in management of river basins that cross jurisdictional boundaries, with

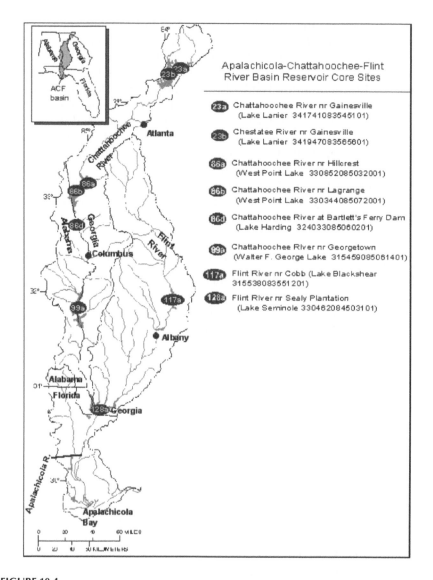

FIGURE 10.4

Map of the Apalachicola–Chattahoochee–Flint (ACF) River Basin. (US Geological Survey: http://ga.water.usgs.gov/nawqa/graphics/ACF.mainstem.gif.)

upstream states typically having different perspectives on water management from those downstream (see Chapter 2 in this volume for a discussion of similar debates around water management in Australia's Murray–Darling Basin). In the ACF River Basin, Georgia, as the primary upstream user, is interested in having enough water supply to continue its growth, particularly in the metropolitan Atlanta area. Alabama, a downstream user, is principally concerned that Atlanta's growth and needs for water will limit its own use for power generation, fisheries, and transportation. Florida, the lowermost downstream user, is concerned with having enough freshwater to reach the Apalachicola Bay to sustain estuaries and coastal fisheries including its culturally and economically significant oyster fishery. In the late 1980s these differences escalated into a heightened state of dispute over shared water resources among the three states. At the time, the region was experiencing drought conditions and steady population growth with increasing water demands. The city of Atlanta and significant portions of the surrounding metropolitan area derive much of their drinking water supply from direct withdrawals and releases from Lake Lanier. Tensions among increasing urban water demands and other water use sectors have continued. Those tensions tend to be reduced during periods of greater rainfall, but are magnified during periods of drought and water shortages, such as during 2006–2009.

In the 1990s the three states began a series of negotiations that ultimately led to an interstate water compact that was ratified by the US Congress in 1997. The compact created a structure that would allow the states to work together to determine a water allocation agreement for the ACF River system while keeping litigation among the states on hold. In 2003 the negotiations ended, however, without the states agreeing to an allocation scheme. As a result of the failed agreement, litigation resumed and remained ongoing at the time of writing (2012).

In July 2009 a US district judge ruled that the US Army Corps of Engineers had illegally reallocated water from Lake Lanier to meet metropolitan Atlanta's needs without approval from the US Congress. The ruling gave the three states 3 years to negotiate a deal and get congressional approval for a compromise that would allow metropolitan Atlanta to continue withdrawing from the lake; otherwise, the Corps of Engineers would be forced to begin releasing water at 1970s levels.

The first stakeholder meetings for the NIDIS ACF pilot started in the fall of 2009 just after this ruling. The degree of mistrust and wariness was evident when one state official refused to sit at a table with officials from the two other states. As the number of NIDIS stakeholder meetings progressed, it became clear that actors in the basin were interested in getting better integration of early warning information but at the same time were unsure what better information would mean for the ongoing litigation. Would enhanced information only perpetuate the conflict by providing better evidence of impacts, for example, from upstream water use?

One of the primary outcomes of the first NIDIS stakeholder meeting in the ACF was to have focused interactions at the sub-basin scale. This translated into holding three separate meetings across the basin to understand specific issues influencing decisions and the priorities for reducing risk to drought in the ACF. At the same time, NIDIS began working with the ACF stakeholders (ACF-S), which is a group of citizens from the basin that began to self-organise in 2009. Recognising that a business-as-usual approach would not be an effective solution to managing water in the basin, the ACF-S assembled a group of stakeholders representing the four major regions and over 50 different sectors and perspectives. The primary mission of the ACF-S has been to achieve an equitable solution for management of the ACF in a way that balances the economic, ecological, and social values across the basin (ACF stakeholders [ACF-S] n.d.).

The ACF-S was initially reluctant to participate in NIDIS and was especially vigilant to avoid the perception that they were too dependent on or complicit in any federal or state activity. As the three sub-basin meetings progressed, however, the ACF-S became more interested in partnering with NIDIS and considered the activity a part of their larger effort to increase dialogue and common understanding across the basin. The ACF-S will likely be an important participant in the development of a regional drought early warning information system in the ACF and their potential role will be described later.

Dividing the ACF into three sub-basins and holding a meeting in each region provided an opportunity to build a network of partners with state agencies, nongovernment organisations, business owners, and universities. In the Colorado example, there was already a large network of institutions to draw on. At least from the federal perspective, the Southeast was more of a challenge because NIDIS was not able to leverage an existing group of federal partners and because some people in the southeast United States are highly suspicious of federal government programs.

At the end of the three sub-basin meetings, a larger full-basin meeting was held to reconcile discussions held around the basin. The timing was fortunate as the end of the meetings coincided with the La Niña that had developed in the fall of 2010. During La Niña the Southeast typically experiences above-average temperature and below-average rainfall during the boreal fall and winter. Winter is also the most important season for the Southeast in terms of recharge for streams, reservoirs, and groundwater. At the full-basin meeting a climate outlook forum was held to discuss the 30-day and 90-day seasonal forecasts and potential implications for drought in the basin. The purpose of the outlook forum was to take advantage of the early warning opportunity and to show how information from federal, state, and local agencies; universities; businesses; NGOs; and concerned citizens could be brought together to provide an assessment of drought and the potential implications if the forecast verified. The outlook forum was successful at bringing people together in the basin—combining multiple types of knowledge and creating common understanding of the status and potential implications of the drought intensifying.

Immediately following the climate outlook forum, a full basin meeting was held to discuss the findings from the three smaller sub-basin meetings. After presenting the major issues of concern and the commonalities across the basin, the participants were asked to prioritise their most important issues related to drought early warning. The participants indicated communication, awareness, and education were the most important attributes of an early warning system. Also ranked high were the production of ACF-focused webinars and outlooks (similar to the climate outlook forum) and better integration and presentation of data. Improvements in forecasts and drought indices and consistency in drought planning among the three states, however, were not highly ranked. The final recommendations were not so much for better information but for improved sharing and flow of information. This is consistent with the findings of the survey of the use of scientific information by Australian farmers reported in Chapter 12.

Starting in early 2011, an ACF drought assessment group that was initiated at the NIDIS full-basin workshop began meeting on approximately a monthly basis via conference calls and webinars. The webinars covered current drought conditions, stream flow and groundwater status, state of ENSO, reservoir elevations, and forecasts for rainfall (5 to 90 days) and stream flow (30 to 90 days). While the three state climatologists participate and provide briefing materials, a consortium of universities called the Southeast Climate Consortium (SECC) has primarily coordinated the assessment group. Given the sensitivities among the states, it was determined that universities in the region would be considered impartial information brokers and could maintain credibility and legitimacy. The SECC is a multidisciplinary group of researchers that focuses on improving capacity and usability of seasonal forecast information and other types of climate-related decision support tools for agricultural, forest, and coastal systems. Since 2011, the ACF drought assessment group has evolved from a small group of individuals to a consistent set of participants and users that includes the state agencies charged with executing drought plans and implementing water restrictions. Most recently, the group has considered issuing collaborative information statements that could be shared with the three states. The production of collaborative statements from the drought assessment group that includes state agency representatives is a welcome advance in the multistate and federal partnership.

The ongoing litigation in the basin and the general mistrust among the actors, however, has been difficult to overcome. Having local groups participate in a process focussed primarily on drought monitoring, though, has allowed people to transcend some of the mistrust while providing a nonthreatening platform for a group of diverse participants slowly to evolve ways to gain confidence, share insight, and integrate information into their decision-making processes. Another factor that has been an important feature of the drought assessment group's success is that drought conditions in the basin have intensified over the development of the pilot. Ongoing drought has helped the group maintain focus and motivated people across the basin to remain in contact.

There have been setbacks as well, however. In September 2011, with the drought assessment group still in its first year, the Georgia state climatologist was replaced. In Georgia, the state climatologist is appointed by the governor; however, there has always been an informal agreement that the position would be based at the University of Georgia. The state climatologist position was moved to the Georgia Environmental Protection Division, which is the state agency responsible for regulating air and water quality, water supply, and hazardous waste, among other activities. The perception of people in the ACF was that this was politically motivated and that the former state climatologist had discussed the possibility of drought and water shortages too frequently. Some actors in the ACF assumed this was an effort to control the messaging around drought, the potential for water restrictions, and, ultimately, the perception of Atlanta not having enough water to sustain its continued growth.

Conclusion

This chapter described two case studies that were developed out of the US National Integrated Drought Information System. The case studies provide insight into the complexity of managing water in federal systems and engaging communities and grassroots organisations in developing early warning systems. They draw attention to the role and importance of existing networks and groups already in place. Depending on the region and the history of groups working together, it may be necessary to focus more energy initially on building cohesive structures and improving dialogue before pushing for linkages between early warning and drought risk-reduction strategies. The most important aspect of this approach is to sustain the interaction long enough so that those linkages can evolve.

In the case of the ACF, the starting point was much different than in the Upper Colorado River Basin example. The advantages of local groups in the ACF creating an inclusive participatory system for communicating early warning information related to drought are clear; however, how the institutions in the region and at the federal level can invent incentives that reward and reinforce cooperation and thus sustain the interaction is another question. Unlike the upper Colorado River Basin example, the US Drought Monitor and the benefits to informing drought plans and providing information for USDA assistance programs have not been a key factor in the ACF.

The politics related to water in the basin remain ever present. The political shadow results from ongoing litigation and political wrangling during the 2006–2009 drought in addition to seemingly overt political decisions such as the replacement of the state climatologist in Georgia, which was widely reported in the media. In this regard the ACF-S could be an invaluable group that eventually could improve and possibly help facilitate the drought assessments and outlooks in partnership with the universities, such as the Southeast Climate Consortium. One of the issues with having universities coordinate the assessment group is that these activities will eventually need to be conducted in an operational setting. The university structure is not always consistent in or conducive to maintaining an operational activity. As the ACF-S develops it is possible they could take the drought assessment activity on and continue the partnership with the universities and federal and state agencies. The ACF-S has gained considerable attention from state agencies, like the Georgia Environmental Protection Division, and from federal agencies like the US Army Corps of Engineers, and others including state congressional delegations simply because they represent a diverse group of interests and are well organised. Given their status, they could be an important group for accountability and helping to mainstream drought early warning information into risk reduction policies and decision making at the state level.

Given the differential levels of cooperation and experience of groups working together, one of the benefits of the NIDIS approach is that it is developing the regional drought early warning information systems as prototypes. Prototyping allows flexibility and acknowledges that getting the correct approach through iteration and learning is important and that mistakes can be made. It also acknowledges that roles do not have to be defined *a priori* and can evolve over time and as experience dictates. As NIDIS learns from the two case studies here and continues developing the RDEWS network elsewhere, it will be important to recognise that interaction with social networks will need to be developed or treated differently depending on the basin and region and the issues of interest.

It should be noted, however, that the power of networks and groups is not always sufficient by itself to overcome all barriers that limit action. There is a risk that improved social interactions can encourage conformity, perpetuate inequity, and allow for the promotion of agendas that have benefits only to certain groups at the expense of others. Despite these potential deficiencies, social networks nevertheless provide an important foundation that must be considered if knowledge is to be accumulated and deployed in the interest of reducing risk associated with drought.

References

ACF stakeholders (ACF-S). n.d. Charters/bylaws. http://acfstakeholders.org/about-acfs/charterbylaws/

Adger, W. N., K. Brown, and E. L. Tompkins. 2005. The political economy of cross-scale networks in resource co-management. *Ecology and Society* 10 (2).

Botterill, L. C., and M. J. Hayes. 2012. Drought triggers and declarations: Science and policy considerations for drought risk management. *Natural Hazards*, 64: 139–151.

Bray, D. B. L., L. Merino-Perez, P. Negreros-Castillo, G. Segura-Warnholtz, J. M. Torres-Rojo, and H. F. M. Vester. 2003. Mexico's community-managed forests as a global model for sustainable landscapes. *Conservation Biology* 17 (3): 672–677.

Cash, D. W., W. N. Adger, F. Berkes, P. Garden, L. Lebel, P. Olsson, L. Pritchard, and O. Young. 2006. Scale and cross-scale dynamics: Governance and information in a multilevel world. *Ecology and Society* 11 (2): Art. 8.

Jodha, N. S. 1990. Common property resources and rural poor in dry regions of India. *Economic and Political Weekly* 21:1169–1181.

National Climatic Data Center (NCDC). 2011. Billion dollar U.S. weather disasters. http://www.ncdc.noaa.gov/oa/reports/billionz.html

National Research Council. 2007. *Colorado River Basin water management: Evaluating and adjusting to hydroclimatic variability.* Washington, DC: The National Academies Press.

Pretty, J., and D. Smith. 2004. Social capital in biodiversity conservation and management. *Conservation Biology* 18 (3): 631–638.

Svoboda, M., D. LeComte, M. Hayes, R. Heim, K. Gleason, J. Angel, B. Rippey, et al. 2002. The Drought Monitor. *Bulletin of the American Meteorological Society* 83 (8): 1181–1190.

Wilhite, D. A. 2000. Drought as a natural hazard: Concepts and definitions. In *Drought: A global assessment,* ed. D. A. Wilhite, 3–18. London: Routledge.

Woodhouse, C. A., S. T. Gray, and D. M. Meko. 2006. Updated streamflow reconstructions for the upper Colorado River Basin. *Water Resources Research* 42:W05415.

11

Ranchers in the United States, Scientific Information, and Drought Risk

Cody L. Knutson and Tonya R. Haigh

CONTENTS

Drought is one of the primary risks faced by cattle and other livestock ranchers in the United States but many ranchers fail to prepare adequately for and respond to drought, resulting in a range of potentially negative social, environmental, and economic effects, as discussed in previous chapters. New drought risk management initiatives in the United States are aimed at producing usable scientific information, management tools, and financial mechanisms to foster proactive management of drought risk among ranchers.

According to the US Forest Service (2011), there are an estimated 770 million acres (311.6 million ha) of rangelands in the United States that account for about 34 percent of the total land area. These rangelands are diverse, from the wet grasslands of Florida to the desert shrub ecosystems of Wyoming, and the high mountain meadows of Utah to the desert floor of California. Private individuals own more than half of these rangelands, the federal government manages 43 percent, and state and local governments manage the remainder.

Despite this climatic, topographical, land use and ownership diversity, one of the common challenges rangeland managers face is the threat of drought and for those dependent on the rangelands for income, the financial impacts can be severe. A survey of cattle producers in the Great Plains states of Nebraska and Texas revealed that severe drought was perceived to be the greatest threat to ranch incomes (Hall et al. 2003). Although research suggests that farmers and ranchers learn from drought events and many perceive that they are becoming more prepared over time (Saarinen 1966; Coppock 2011), a number of barriers have been identified that limit ranchers' ability to prepare adequately for and respond to drought events.

Barriers to Drought Risk Management

An oft cited key barrier to drought risk management is a lack of awareness of the drought hazard. Studies have shown that Great Plains farmers tend to have a selective memory when recalling past drought events, as anticipated in the discussion of attitudes to risk in Chapter 2. They tend to remember the first, the worst, and the most recent droughts they experience (Saarinen 1966; Taylor et al. 1987; Woudenberg et al. 2008). In addition to producing inaccurate recollections of drought frequency, the range of drought conditions experienced can help form a mental model of how drought risk should be managed (Taylor et al. 1987; Dunn et al. 2005). Therefore, rangeland managers who lack drought experience or who have had a recent run of good years may have less awareness of drought as a hazard and the full range of potential effects. This is important as past negative experiences and awareness of recurrent drought can lead to changes in response patterns (O'Connor et al. 2005; Coppock 2011).

Similarly, Thurow and Taylor (1999) pointed out that the identification of drought is often difficult for ranchers (and others), which frequently results in lagged management responses. The US National Drought Policy Commission (2000) also reported that ranchers may lack information about local drought conditions and outlooks, which can limit their ability to make timely drought risk management decisions (prior to and during drought). Several studies have shown that ranchers around the world do not have a good deal of confidence in climate outlooks, which limits their use in decision making (Ash et al. 2000; Jochec et al. 2001; Luseno et al. 2003; Coppock 2011). A lack of confidence in outlook accuracy and reliability and outlooks that are too geographically broad are among the most cited criticisms.

Even if rangeland managers do perceive a drought threat, there are other management barriers to implementing drought risk reduction actions. Knowing the best way to prepare for and respond to drought is a difficult task, and scientific evaluations to assist in determining the optimal application of drought-related measures for a ranching operation are lacking (Bastian et al. 2006). In a recent study by Coppock (2011), ranchers were asked what could help them become better drought managers. A frequently cited suggestion was to improve their access to information concerning best practices for drought management. Determining appropriate measures is complicated by the fact that the ranching industry and related systems are changing over time and some traditional drought risk reduction strategies may no longer be as viable as they once were. For example, Dunn et al. (2005) describe how it was once common practice for a ranch to have both a cow and yearling cattle herd. The yearlings could graze excess forage in years when it was abundant and be sold quickly during drought years. For several reasons, however, the supply of desirable yearlings that ranchers could utilise is now smaller and ranchers are instead maintaining larger cow herds, which they may be hesitant to liquidate, even in dry periods. Such developments in ranch management, along with the limited access to information about potential risk reduction strategies and their financial effects, can exacerbate tendencies to uncertainty and inaction (Dunn et al. 2005).

It has also been argued that technical and financial assistance for the development of ranch-level drought plans is lacking in the United States. The US National Drought Policy Commission (2000) noted that less than 10 percent of US farmers and ranchers were receiving technical assistance to help them develop and implement drought plans, and an even smaller number were receiving cost-share assistance for planning. At the time, the commission also stated that federal drought mitigation and response programmes for livestock producers were also lacking. For example, during listening sessions held by the commission, livestock producers consistently pointed out that federal insurance was not available to help reimburse losses and that their operations were typically excluded from emergency relief assistance.

The issue of drought-related assistance from the government is a contentious issue. Since the devastating droughts of the 1930s (the Dust Bowl years), a variety of government assistance has been provided to ranchers to help prepare for, respond to, and recover from drought. Although it is recognised that some forms of assistance are beneficial for helping ranch operations becoming more resilient to drought (for example, training and resource development cost-share programmes), other drought relief programmes have been criticised for encouraging livestock owners to delay appropriate drought response strategies (Thurow and Taylor 1999; Dunn et al. 2005). Some of these, notably emergency relief policies, have been deemed to invoke 'moral hazard' and encourage riskier behaviour because the participants know they will be buffered from negative consequences and are less likely to implement timely risk reduction measures proactively (Thurow and Taylor 1999). On the other hand, Coppock (2011, 616) argues that since insurance payments and government relief only partially compensate drought-related production losses, and relief assistance is often unpredictable, 'It makes little sense that a rancher would gamble on full access to generous government assistance during or after drought'. Although Coppock felt drought relief programmes may have made a difference to some ranchers in coping with drought, such programmes do not fundamentally influence rancher decision making. These types of debates illustrate the contention that is evident in trying to determine the appropriate role of the government in helping ranchers reduce drought risk.

Overall, several barriers have been identified that limit drought risk reduction among ranchers in the United States, such as misperceptions of past drought, a lack of awareness of or access to current drought monitoring and outlook information, limited confidence in the accuracy and reliability of drought monitoring and outlook information, uncertainty in regard to potential management options and their financial outcomes, unavailability of drought insurance products, and potential disincentives for proactive management as a result of government relief programmes. During the past two decades, attention has increasingly been focussed on trying to understand and overcome these barriers to enhance drought risk reduction in the ranching sector.

Increased Emphasis on Drought Risk Management for Ranching Operations

Although there are continual efforts by individual ranchers and advisors to develop ranching systems that are better adapted to local conditions, several factors have culminated in a recent increase in the emphasis on drought risk management. First, the recurrence of especially severe droughts (for example, 1987–1989, 1995–1996, 2002–2006) has demonstrated the continued vulnerability of the ranching industry to drought. As with any drought, these events focussed attention on what could be done to help vulnerable sectors, including ranching operations, better prepare and respond to similar occurrences in the future. This comes at a time when the subject of drought risk management has gained prominence worldwide (United Nations Strategy for Disaster Reduction [UNISDR] 2009).

In the United States, this attention has resulted in a number of articles and guidance documents by ranching advisors that stress drought risk management as an essential component of management planning (Reece et al. 1991; Thurow and Taylor 1999; Pratt 2000; Hamilton 2003; Hart and Carpenter 2004; National Drought Mitigation Center (NDMC) 2011). The run of severe droughts also gained the attention of national policy makers and funding agencies, providing the impetus to support new forms of assistance to address more proactively drought-related issues in the United States. For example, the NDMC was formed at the University of Nebraska-Lincoln in 1995 with the goal of providing assistance to reduce societal vulnerability to drought, the National Drought Policy Act of 1998 was passed by the US Congress to authorise an investigation of how the country could better prepare for and respond to drought, and the National Integrated Drought Information System Act of 2006 was passed by the US Congress to foster collaboration on the development of enhanced drought early warning services (see Chapter 10).

The National Drought Policy Act of 1998, in particular, encouraged greater attention to the drought-related needs of the ranching industry. As outlined in Chapter 10, to investigate drought risk management in the United States, the act authorised the creation of the National Drought Policy Commission (NDPC), which made several recommendations directly related to livestock production. As reported by the NDPC (2000), during the investigation, many comments were received stressing the importance of moving away from the traditional approach to drought that is driven by emergency relief to a new approach that emphasises planning and proactive mitigation, a trend also discussed in relation to Australian policy in Chapters 4 and 6. However, in a compromise, the Commission's final recommendations emphasised assistance to help ranchers mitigate and plan for drought, while at the same time maintaining a safety net of emergency relief. One of their guiding conclusions was that the United States should 'favor preparedness over insurance, insurance over relief, and incentives over regulation'. Specifically in regard to agriculture, the Commission recommended that the US Congress adequately fund existing drought preparedness programmes in the US Department of Agriculture and authorise and fund the department to

1. Cooperate with state and local governments and the private sector to expand training to rural communities, farmers, and ranchers across the country on various financial drought management strategies
2. Evaluate different approaches to providing crop insurance to stakeholders, including ranchers
3. Amend their emergency programmes to include livestock needs during drought

In addition, the Commission recommended actions to improve collaboration among scientists and decision makers to enhance drought monitoring, prediction, information delivery, and research to foster public understanding of and preparedness for drought. They also recommended that the President direct Congress to fund existing and future drought-related research adequately, and that competitive research grant programmes give a high priority to drought. Although these were only recommendations, they did shed light on drought-related issues faced by ranchers and other stakeholders and provided justification for placing more attention on drought risk reduction in the United States.

This need for drought risk reduction was reinforced by the occurrence of extreme drought across much of the western portion of the country during the early 2000s (peaking in 2002), where the widespread effects of the drought on ranching operations 'may have garnered strength for changes in federal legislation' (Tronstad and Fuez 2002). For example, passage of the

National Integrated Drought Information System (NIDIS) Act of 2006 was undoubtedly influenced by the effects of these drought conditions. This Act authorised the creation of a NIDIS Program Office within the US National Oceanic and Atmospheric Administration, which has been a champion for fostering collaboration and education on drought-related issues in the United States at the federal level.

These factors have helped provide the momentum, political authority, and resources to increase the focus on drought risk management and planning, resulting in the development of a range of new scientific information, tools, and programmes to help ranchers and other stakeholders understand and address drought risk.

New Tools to Assist Ranchers

During the last two decades, many of the activities implemented to assist ranchers have focussed on developing new scientific information and tools to help them make more informed management decisions, as well as investigating methods for their effective dissemination. Other activities have included the creation of new insurance products and relief measures to help ranchers during drought, many of which utilise new drought tools in their implementation.

Drought Monitoring Activities

Drought monitoring is one research area that has experienced significant recent growth in the United States. During the 1995–1996 drought that affected the Southwest and southern Great Plains states, researchers at the newly established NDMC began investigating if there was a better way to define and track drought conditions on a consistent basis across the United States. According to Svoboda (2000), at that time there were no comprehensive monitoring activities being conducted by or between various federal, state, or regional entities. Enhanced drought monitoring was thought to be one way to help foster a better understanding of the phenomenon, raise awareness, and foster more effective and timely drought responses. Passage of the 1998 National Drought Policy Act encouraged federal agencies to collaborate with the NDMC in creating the US Drought Monitor, which was first published online in 1999 and has since provided a weekly snapshot of general drought conditions around the United States using a 'convergence of evidence' approach to produce the composite index of drought (M. Svoboda, NDMC, pers. comm., October 25, 2011). When the Drought Monitor was created, authors had relatively few drought indicators that were consistently provided in helping them to form a weekly map. The creation of the monitor and the resulting dialogue among the climate research community has fostered the creation of many new monitoring indicators to fill information gaps. As a result, Drought Monitor authors now have approximately twenty-five to thirty different indicators that are evaluated on a weekly basis to produce the drought map (Svoboda 2011) and there are increasing efforts to provide even more localised data for decision support, including those useful for ranchers.

One of these new indicators is the Vegetation Drought Response Index[*] (VegDRI) created jointly by the NDMC, the US Geological Survey's National Center for Earth Resources Observation and Science, and the High Plains Regional Climate Center. The tool, released in 2006 and funded largely by the USDA Risk Management Agency (RMA), uses a variety of climatic and biophysical variables to produce a weekly map of vegetation stress due to drought on crop and rangelands across the lower 48 US states. A benefit of VegDRI for ranchers is that it provides relatively high-resolution data on current vegetation conditions attributable to drought, which can be useful for understanding local range conditions and comparing conditions with other regions and, eventually, past events.

The NDMC also released the US Drought Impact Reporter[†] (DIR) in 2005, which is the nation's first comprehensive database of impacts. The DIR is populated with news stories and reports on a daily basis to track drought impacts across the country, and it includes an option for ranchers and other affected persons to submit reports of local drought conditions. The tool was developed with financial support from the US National Oceanic and Atmospheric Administration and the USDA Risk Management Agency. The Drought Impact Reporter serves as an important archive of drought impacts, allows for tracking and comparing impacts across the country over time, and provides a way to ground-truth national monitoring assessments with local conditions.

Another new scientific tool is the US Seasonal Drought Outlook, which was first released by the US Climate Prediction Center in 2003.[‡] The 90-day outlook is expressed as a map showing areas of the country where drought conditions are expected to develop, worsen, stay the same, or improve. The goal is to highlight regions at risk so that users can take proactive measures to mitigate the impacts of and respond to drought.

To help foster improved access to these types of drought information, the NIDIS Program Office also created the US Drought Portal.[§] The portal provides access to a variety of information on drought monitoring, impacts, planning, and education. These types of new tools have the overall goal of providing enhanced access to better information about the occurrence of drought, which

---

[*] www.vegdri.unl.edu
[†] www.droughtreporter.unl.edu
[‡] http://www.cpc.ncep.noaa.gov/products/expert_assessment/seasonal_drought.html
[§] drought.gov

is expected to lead to a better understanding of the drought hazard and foster more proactive management by stakeholders such as ranchers.

Drought Risk Management on the Ranch

To help promote self-reliance and sustainability at the individual level, the federal government has also funded the development of a number of new tools to help producers make management decisions at the ranch level. For example, the USDA Agricultural Research Service and USDA Natural Resources Conservation Service (NRCS) have supported the development of various 'drought calculators' for livestock producers in the states of North Dakota and South Dakota. The calculators, though slightly different, are each built upon scientific knowledge of the relationship between precipitation amount and timing, and forage production. The North Dakota tool, the *Drought Management Calculator,* was developed by a consortium of USDA and university researchers and funded by the USDA Risk Management Agency.* The calculator uses monthly precipitation data and the producer's livestock data to predict ideal stocking rates and feed needs. The programme also integrates market prices and feed costs into recommendations and reports. After the producer creates a report, the tool can be used to develop several 'if–then' scenarios to support decision making. On the other hand, the *South Dakota Drought Tool*† developed by the USDA NRCS uses only precipitation data, and if drought conditions warrant, provides basic drought management recommendations.

In addition, new software such as *The Grazing Manager* was developed by researchers at Texas A&M University and a private consulting company (Agren, Inc.) with funding from the USDA RMA (released in 2004) to assist producers in monitoring forage production and stocking rates.‡ This type of software allows users to track growing conditions, livestock numbers, grazing days by pasture, and forage utilisation to set goals, track performance, and test and improve management decisions. The Grazing Manager is meant to simplify 'the complex decision environment and gives users the ability to test decisions through simulation, respond to critical environmental changes, and more effectively manage towards meeting goals' (Agren, Inc. 2008, 4). It helps users address drought as they see conditions significantly deviating from their 'normal' situation.

To help provide improved access to information about drought and its effects on ranch production, drought risk management strategies and tools, and planning information, the NDMC released the Managing Drought Risk on the Ranch website in 2011 with funding from the USDA Risk Management

---

* http://www.nd.nrcs.usda.gov/technical/Drought_Management_Calculator.html
† http://www.sd.nrcs.usda.gov/technical/Technical_Tools.html
‡ http://www.thegrazingmanager.com/drupal

Agency.* The NDMC worked with ranchers, advisors, and researchers to develop the web-based guide, which includes steps common to risk management planning (that is, inventory of resources, identification of risks, developing strategies to manage drought risk, and monitoring ongoing performance). The NDMC's guide helps users identify critical times for making decisions. This ranch drought-planning model advocates critical dates and triggers as a science-based means of decision making. The use of critical dates and triggers helps producers overcome specific barriers to action, including the complexity of choice and perception of control over the situation.

The goal of these types of activities is to help ranchers incorporate the best available information into their ranch operation decision making by increasing their knowledge about drought and its effects on their operation, providing them with new tools to help them make decisions, and enhancing access to the information and tools.

Government Assistance Programmes

Recurrent drought and the increased emphasis placed on meeting the needs of ranchers also resulted in a range of new government assistance programmes, many of which are increasingly using new drought monitoring tools to trigger their activation. For example, the federal government responded to the 2002 drought by passing the Agricultural Assistance Act of 2003, which authorised the Livestock Assistance Program that made funds available to livestock producers in disaster areas who suffered grazing losses in either 2001 or 2002. Similarly, in 2003, the USDA established the Non-fat Dry Milk Livestock Feed Assistance Program to provide surplus USDA stocks of non-fat dry milk to livestock producers for supplemental feed in areas hardest hit by continuing drought. The newly created US Drought Monitor was used to determine which counties were eligible to ensure that the initiative was targeted to producers in greatest need (US Department of Agriculture Farm Services Agency [FSA] 2003).

The federal government has followed these measures with a variety of assistance programmes in subsequent years. For example, the Food, Conservation, and Energy Act of 2008 (2008 Farm Bill) amended the Trade Act of 1974 to create supplemental agricultural disaster assistance programmes to replace previous disaster programmes, although the former expired September 30, 2011, and had not been renewed at the time of this publication. Again, assistance through these programmes was triggered by drought status as indicated in the US Drought Monitor (US Department of Agriculture Farm Services Agency 2009; Monke 2010).

One of the supplemental agriculture disaster assistance programmes, the Livestock Forage Disaster Program (LFDP), required that producers purchase Federal Crop Insurance Corporation insurance through the

---

* drought.unl.edu/ranchplan

Non-insured Assistance Program or the Pasture, Rangeland, Forage (PRF) plan of insurance. The PRF insurance programme was introduced in 2007 as a pilot programme in portions of nine states (Colorado, Idaho, North Dakota, Oklahoma, Oregon, Pennsylvania, South Carolina, South Dakota, and Texas) to provide income when rangeland, pasture, and forage production is negatively impacted (US Department of Agriculture Federal Crop Insurance Corporation Risk Management Agency [RMA] 2007). As of 2011, it had been expanded to twenty-four states, including most of the western United States (Pugh 2010). The PRF plan of insurance is a group risk plan, intended to provide protection against widespread loss of production based on either gridded average precipitation or vegetation greenness, depending upon the state and county. The programme provides indemnities based on an overall index, not individual losses. The rainfall index used in the programme is based on NOAA rainfall data and the vegetation index is based on the Normalized Vegetation Deviation Index produced by the US Geological Survey Earth Resources Observation and Science Center (RMA 2009).

Because this is a new programme, the jury is still out on its impacts on long-term rangeland sustainability. According to the RMA (2006), because losses are based on gridded averages, individual losses resulting from poor management are not 'rewarded'. However, other researchers who have looked at rainfall-index insurance in the context of developing countries have cautioned that such insurance with frequent payoffs could lead to less sustainable grazing management in the long term by enabling producers to overstock pastures during drought (Muller et al. 2009; Bhattacharya and Osgood 2008).

New Drought Planning Incentives for Livestock Producers

Policy makers and advisors in the United States have long debated how to best assist ranchers in preparing for and responding to drought. The US National Drought Policy Commission (2000) promoted an approach of self-reliance and incentives supplemented with insurance and emergency relief programmes. Thurow and Taylor (1999) have noted the potential benefits of mandating that ranchers create and implement drought response plans, which would be approved by the federal government prior to receiving certain emergency assistance programmes in order to resolve the 'moral hazard' issue.

Mandating the development of drought plans by ranchers has not been adopted on a national scale in the United States. However, the USDA Natural Resources Conservation Service (NRCS) in South Dakota recently piloted a project that required drought plans. The programme, called the Grazing Lands Sustainability Initiative and offered through the Environmental Quality Incentives Program, was offered from 2008 to 2011 to improve drought sustainability on grasslands. The programme provided financial incentives to encourage rangeland management that helped producers prepare for drought. The programme required that participating producers develop a written drought contingency plan, with the assistance of NRCS if needed, and attend a grazing school to learn monitoring, grazing management, and planning. According to NRCS Range Specialist Stanley Boltz (pers. comm. 2011), between ten and thirty-five producers participated in the programme each year during the programme's existence, and responses to participation in the grazing school were overwhelmingly positive. Environmental quality incentives programmes throughout the United States may still emphasise the need for contingency planning, and some include participation in a grazing school in their ranking criteria, but the drought-planning emphasis of the pilot Grazing Lands Sustainability Initiative has, thus far, not been replicated.

Conclusion

Ranchers will always be faced with uncertainty (imperfect knowledge) when dealing with inherent climatic variability, market prices, and external financial considerations (Thurow and Taylor 1999). However, new scientific information and tools (for example, the US Drought Monitor, Vegetation Drought Response Index, Drought Impact Reporter, and US Seasonal Drought Outlook) have been developed to help provide them with a better understanding of historical, current, and potential drought occurrence, including what constitutes a drought. Despite limitations of this information, such as the accuracy and reliability of outlooks, they do provide additional information that ranchers can use as inputs in their decision making.

A great deal of investment has also focussed on the development of new information and tools that will help ranchers make decisions in their operation to reduce risk (uncertain consequences) associated with climatic variability. New 'drought calculators', grazing software, and drought planning guides have been developed to take some of the guesswork out of ranching. As discussed by Stuth et al. (1991), such tools are intended to foster a transition of management from a heuristic art to an analytical decision-making process. In explanation, they go on to state:

> Scientifically defensible knowledge-based decision support systems offer the manager a means of blending his knowledge with that of experts in a planning environment which he controls. He can access external information and develop 'what if' scenarios reflecting his personal perception of the current situation and future responses. Once in control, the value of external information can better be assessed and personal perceptions tested. This process allows for technology to be evaluated in relation to unique planning environments.

The use of these types of science-based tools has great potential for reducing uncertainty in decision making and enhancing the development of more informed drought plans.

Similarly, new drought assistance programmes have been implemented to assist ranchers in preparing for and responding to drought during the last decade. In order to be applied equitably, these policies are increasingly tied to science-based products to determine where and when they should be implemented. The US Drought Monitor is one example of a source of information that is increasingly used to define drought and trigger drought assistance by several government organisations. The rainfall index used as the basis for triggering relief payments through federal insurance programmes is another example. Although their application has limitations (for example, the resolution of the US Drought Monitor is not typically at the scale where declarations are often made), they do represent the use of best available scientific information.

However, new information and tools are not beneficial if they are not accessible to potential users. Therefore, a great deal of investment is also being placed on their dissemination through outlets such as websites of the National Integrated Drought Information System, the National Drought Mitigation Center, and other relevant entities. Emphasis has also been placed on providing technical assistance and incentives for implementing drought risk reduction practices at the ranch level through programmes like the USDA NRCS pilot programme in South Dakota.

Despite these efforts, additional work is needed on several fronts. Developing localised drought monitoring information and accurate climate outlooks suitable for ranch-level planning, identifying optimal drought risk management strategies, providing access to new information and tools, and determining the appropriate motivations to encourage ranchers to use the information are still all significant challenges in reducing drought risk. However, significant steps are being taken to develop the scientific information, tools, and policies necessary to foster a more balanced approach to drought risk management in the United States that fosters self-reliance through appropriate government intervention.

References

Agren, Inc. 2008. *The grazing manager user guide.* http://www.thegrazingmanager.com/drupal/TGMfiles/TGM_user_guide.pdf

Ash, A., P. O'Reagain, G. McKeon, and M. Stafford Smith. 2000. Managing climate variability in grazing enterprises: A case study of Dalrymple shire, north-eastern Australia. In *Applications of seasonal climate forecasting in agricultural and natural ecosystems: The Australian experience,* ed. G. L. Hammer, N. Nicholls, and C. Mitchell, 253–270. Dordrecht: Kluwer Academic.

Bastian, C., S. Mooney, A. Nagler, J. Hewlett, S. Paisley, M. Smith, W. M. Frasier, and W. Umberger. 2006. Ranchers diverse in their drought management strategies. *Western Economics Forum* 5 (2): 1–8.

Bhattacharya, H., and D. Osgood. 2008. Index insurance and common property pastures. Department of Economics working papers, http://www.econ.utah.edu

Coppock, D. L. 2011. Ranching and multi-year droughts in Utah: Production impact preparedness. *Rangeland Ecology and Management* 64 (6): 607–618.

Dunn, B., A. Smart, and R. Gates. 2005. Barriers to successful drought management: Why do some ranchers fail to take action? *Rangelands* 27 (2): 13–16.

Hall, D. C., T. O. Knight, K. H. Coble, A. E. Baquet, and G. F. Patrick. 2003. Analysis of beef producers' risk management perceptions and desire for further risk management education. *Review of Agricultural Economics* 25 (20): 430–448.

Hamilton, W. T. 2003. Drought: Managing for and during the bad years. In *Ranch management: Integrating cattle, wildlife, and range*, ed. C. A. Forgason, F. C. Bryant, and P. C. Genho, 133–152. Kingsville, TX: Kingsville Ranch Institute.

Hart, C. R., and B. B. Carpenter. 2004. *Planning: The key to surviving the drought.* AgriLIFE Extension: Texas A&M University.

Jochec, K. G., J. W. Mjelde, A. C. Lee, and J. R. Conner. 2001. Use of seasonal climate forecasts in rangeland-based livestock operations in west Texas. *Journal of Applied Meteorology* 40:1629–1639.

Luseno, W. K., J. G. McPeak, C. B. Barrett, P. D. Little, and G. Gebru. 2003. Assessing the value of climate forecast information for pastoralists: Evidence from southern Ethiopia and northern Kenya. *World Development* 31:1477–1494.

Monke, J. 2010. *Previewing the next farm bill: Unfunded and early expiring provisions.* Congressional Research Service. http://www.crs.gov

Muller, B., M. Quaas, K. Frank, and S. Baumgartner. 2009. Pitfalls and potential of institutional change: Rain-index insurance and the sustainability of rangeland management. University of Lüneburg working paper series in economics.

National Drought Mitigation Center (NDMC). 2011. Managing drought risk on the ranch 2011 [Accessed October 17, 2011. Available from http://drought.unl.edu/ranchplan].

O'Connor, R. E., B. Yarnal, K. Dow, C. Jocoy, and G. Carbone. 2005. Feeling at risk matters: Water managers and the decision to use forecasts. *Risk Analysis* 25 (5): 1265–1275.

Pratt, D. W. 2000. Drought proofing your business. *Ranching for Profit Newsletter* #65: Ranch Management Consultants.

Pugh, S. 2010. RMA expands pasture, rangeland, forage insurance availability for 2011 crop year. US Department of Agriculture Risk Management Agency News. http://www.rma.usda.gov/news/2010/06/prfexpansion.html

Reece, P., J. D. Alexander, and J. R. Johnson. 1991. *Drought management on range and pastureland.* Nebraska Cooperative Extension.

Saarinen, T. E. 1966. *Perception of the drought hazard on the Great Plains.* Chicago: University of Chicago Press.

Stuth, J. W., J. R. Connor, and R. K. Heitschmidt. 1991. The decision-making environment and planning paradigm. In *Grazing management: An ecological perspective*, ed. R. K. Heitschmidt and J. W. Stuth 201–233. Portland, OR: Timber Press.

Svoboda, M. 2000. An introduction to the Drought Monitor. *Drought Network News* 12 (1): 15–20.

———. 2011. National Drought Mitigation Center, October 25, 2011.

Taylor, J., T. Stewart, and M. Downton. 1987. Perceptions of drought in the Ogallala Aquifer region of the Western U.S. Great Plains. In *Planning for drought: Toward a reduction of society vulnerability,* ed. D. Wilhite, W. Easterling, and D. Wood, 409–423. Boulder, CO: Westview Press.

Thurow, T., and C. Taylor. 1999. The role of drought in range management. *Journal of Range Management* 52 (5): 413–419.

Tronstad, R., and D. Fuez. 2002. Impacts of the 2002 drought on western ranches and public land policies. *Western Economics Forum* 1 (2): 19–24.

United Nations Strategy for Disaster Reduction (UNISDR). 2009. *Drought risk reduction framework and practices: Contributing to the implementation of the hyogo framework for action.* Geneva, Switzerland: United Nations Secretariat of the International Strategy for Disaster Reduction.

US Department of Agriculture Farm Services Agency (FSA). 2003. *Surplus non-fat dry milk sales for feed program.* [Accessed October 26, 2011. Available from http://fsa. usda.gov/Internet/FSA_File/nfdmqa.pdf].

———. 2009. *Livestock forage disaster program.* http://www.fsa.usda.gov/Internet/ FSA_File/lfp09.pdf

US Department of Agriculture Federal Crop Insurance Corporation Risk Management Agency (RMA). 2006. *Pasture, range, forage (PRF) plans of insurance.* [Accessed October 25, 2011. Available from http://www.rma.usda.gov/policies/pasture-rangeforage/PRF-Combined-Condensed.pdf].

———. 2007. *Pasture, rangeland, forage pilot programs—2007.* http://www.rma.usda. gov/policies/pasturerangeforage/2007availabilitymap.pdf

———. 2009. *Rainfall index insurance standards handbook 2010 and succeeding crop years.* http://www.rma.usda.gov/handbooks/18000/2010/10_18130.pdf

US Forest Service. 2011. About rangelands. http://www.fs.fed.us/rangelands/ whoweare/index.shtml

US National Drought Policy Commission. 2000. *Preparing for drought in the 21st century.* Available from http://govinfo.library.unt.edu/drought

Woudenberg, D. L., D. A. Wilhite, and M. J. Hayes. 2008. Perceptions of drought hazard and its sociological impacts in south-central Nebraska. *Great Plains Research* 18:93–102.

12

Drought Science and Policy: The Perspectives of Australian Farmers

Geoff Cockfield

CONTENTS

This chapter is a summary and discussion of the results of a survey of Australian farmers, as a sample of potential drought science and policy end users, conducted in May 2012. There were four main areas of investigation developed from the discussions and contentions in other chapters of this volume:

- End users' perceptions of drought as a hazard, as discussed in Chapter 2, and how they see climate change affecting drought (Chapters 1, 2, 3, and 4)
- Access to and use of scientific and policy information relating to drought, which was introduced in Chapters 1 and 2 and more extensively considered by some of the scientists engaged in knowledge brokering (Chapters 7, 10, and 11)
- Drought preparation strategies and the roles of science associated with that, which were discussed in Chapters 7, 10, and 11
- Perceptions of what factors influenced governments in developing drought policy and what sorts of policies survey participants saw as useful

The latter two areas relate to an overarching theme developed across several chapters (1, 2, 4, 5, and 6), which is that, in Australia and the United States, governments have wanted to move end users and government policies from being predominantly reactive to a greater focus on preparation for 'natural' events.

From this survey, the respondents overwhelmingly see drought as a natural event and the majority do not think that climate change will increase the number of droughts, somewhat in contrast to the concerns of some climate scientists. The sample end users appear to be generally relying on basic meteorological information to anticipate seasonal conditions and there is little use of the various drought management and decision-support packages that have been developed over the last 20 years in Australia. Respondents regard 'adaptive' strategies, such as reducing livestock numbers as conditions worsen, as important responses to drought, which is an understandably pragmatic approach, given the current state of forecasting. In regard to policy, there is strong support for governments to cushion the impacts of drought through various forms of support, including some forms that are little used in Australia anymore. This suggests that despite more than 30 years of admittedly spasmodic attempts to convince farmers of the need for self-sufficiency, attitudes may not have shifted significantly and that, once benefits have been given, there is a lingering attachment to these. The chapter concludes with some cautious conclusions, the caution being warranted by the limitations of the survey. This is an exploratory survey to bring together some of the main points from this volume through some end user responses.

Survey Development, Method, and Sample

It was argued in Chapter 2 that attitudes to risk were influenced by whether or not the hazard was seen as natural (Slovic 1987, 281; Renn 2008, 109), so respondents were asked for their perceptions in relation to that and also asked about the potential for climate change to cause an increase in the frequency of droughts, as some climate scientists expect for parts of Australia (Hennessy et al. 2008; Kirono, Kent, et al. 2011; Kirono, Hennessy, et al. 2011). Those involved in communicating drought science and preparedness strategies who contributed to this volume were variously optimistic to cautious about the prospects of influencing end users to be more prepared and to use systematic approaches to management. Stone and Cockfield (Chapter 7) described the difficulties in operating in the triangular relationships of scientists, policy makers, and end users, while McNutt et al. (Chapter 10) and Knutson and Haigh (Chapter 11) are optimistic about the US drought management strategies in development.

For this survey, in order to develop something of a baseline of current practices, respondents were asked about the importance of various drought management strategies. Then they were asked about how they derived information that could enable them to anticipate seasonal conditions and the usefulness of the various sources and types of information. Respondents were then asked about what government programmes and policy instruments were considered most useful and under what circumstances governments should assist farmers. This was to explore current attitudes to different forms of assistance, especially considering the reactive and risk-sharing policies, whereby farmers receive financial assistance and management advice during the drought, as well as assistance and incentives to encourage preparation for drought.

We commissioned a telephone survey of farmers, distributed across all Australian states for a final sample population of 390. Farm enterprises are classified according to the predominant commodity or commodities and the categories include grain production, grain and livestock, beef and sheep, sheep, dairy, and sugar. The survey excludes major irrigated crops such as cotton and rice. Respondents were also classified according to proportion of off-farm income (none; 1–49 percent; 50+ percent), level of annual income, self-assessment of viability (unviable, just viable, easily viable, don't know), age, and gender. The respondents were almost 90 percent male and 61 percent were aged 55 or more, with almost 28 percent aged 65 or more. This apparently skewed sample is a common feature of surveys of Australian farmers, especially when seeking the main business decision maker, so the sample warrants some of the caution expressed at the conclusion of this chapter. Chi-square statistical tests were used to identify differences in responses by any of the variables; only those relationships that are statistically significant are reported.

Results

The respondents overwhelmingly see long periods of dry weather as natural events and the majority do not think climate change will lead to an increase in the number of droughts (Figure 12.1). There were no significant variations by location (state), predominant commodity production, farm size, or age except that with the latter variable those aged 65 years or more were more likely to disagree strongly that climate change would increase the frequency of droughts. More than 24 percent of respondents are not sure of the effects of climate change on drought, so there is also a group presumably open to further evidence.

Respondents rely on a conventional array of indicators of the severity of drought (Table 12.1): length of time without rain and low moisture and water levels. There is some variation by enterprise; dairy farmers consider feed availability to be more important as an indicator, while the beef and sheep producers rate ground cover more highly. For anticipating seasonal conditions, most rely on (short-term) rainfall forecasts and the medium-term

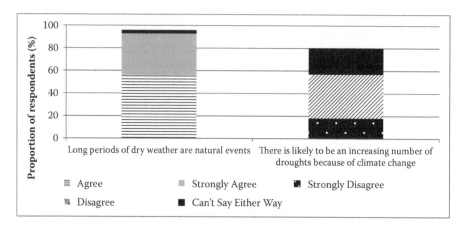

FIGURE 12.1
Perceptions of drought and the potential impact of climate change.

outlook from the Southern Oscillation Index (SOI), which can foreshadow an El Nino event, which is associated with lower rainfall for Australia. Surprisingly, there is also considerable attention to the Indian Ocean Dipole, which is a comparatively recent focus of research. The Western Australian farmers are more likely to use this indicator, since the underlying climatic effect relates most to the ocean adjacent to that state, but there is still considerable use of it elsewhere.

In terms of 'usefulness' for anticipating seasonal conditions, it is the short-term forecast that is considered most useful, though only by 37 percent of respondents. The potential medium-term forecasts (SOI and Dipole) have much less support as being the most useful. There is little use of their own or others' 'experience' or 'common sense'—a question that was included to see

TABLE 12.1

Use and Usefulness of Indicators of Seasonal Conditions

	Proportion of Respondents (%)	
	Which types of information do you use to anticipate seasonal conditions?	**Which type of information is the most useful to you in anticipating seasonal conditions?**
Rainfall received	74	35
Rainfall forecasts	62	24
Southern Oscillation Index	54	18
Soil moisture information	39	5
Groundwater availability	38	4
Indian Ocean Dipole	34	5
Experience/knowledge/ common sense	3	2

TABLE 12.2

Perception of Scientific Information Gaps

What additional scientific information would assist you in managing drought?	Proportion of Respondents (%)
Don't know/none	39
More accurate forecasts	29
Long term forecasts	12
Agronomic advice	6
Easy to understand forecasting	5
Experience/knowledge/personal education	5
Government assistance	4
Economic conditions/commodity prices	4
Technology innovation/R&D	3
Water availability	3
Historical weather data	3

if there is still some reliance on accumulated experience or local theories and knowledge of rainfall patterns. It was noted in Chapter 7 that communities and farmers may be resistant to external information and advice if this conflicts with their local traditional indicators (see also Patt and Gwata 2002). In this survey, local observations are used more for monitoring seasonal conditions than for anticipating the season (Table 12.2).

The main sources of information for monitoring and anticipation of seasonal conditions are mainstream sources including television, websites, and local observations. The main websites are those of the Bureau of Meteorology and Elders, a major agribusiness corporation. The information that is used is considered reasonably reliable and timely (Figure 12.2), though warnings of a forthcoming drought are considered less reliable. Even so, more than half the respondents think that drought warning information is of at least 'average' quality, while 20 percent think the information is good to excellent.

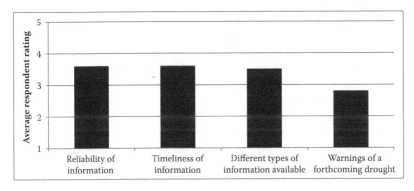

FIGURE 12.2

Opinions of the reliability and timeliness of drought information.

Approximately 63 percent of grain farmers and 62 percent of farmers from Western Australia (WA) think that the reliability of warnings of a forthcoming drought is poor or very poor compared with 35 percent for the whole sample. Grain farmers may have less flexibility in adapting to the onset of drought, given relatively fixed growing seasons, while WA has experienced a series of very dry years and was one of the few areas, along with northern Victoria, with relatively low seasonal rainfall at the time of the survey.

The respondents see conventional drought management strategies—such as storing fodder, reducing stock numbers as it gets drier, and using income smoothing taxation instruments in the form of farm management deposits (FMDs) (see Chapter 6)—as important. There are variations by enterprise, so, for example, dairy farmers are less supportive of selling off livestock— understandably given the importance of breeding for production—and grain farmers see reduced tillage as important. For the FMDs, farmers are able to put money into the deposits in high income years and withdraw these in lower income years, thereby reducing the tax liability and reducing income peaks and troughs. Those who see themselves as unviable rate FMDs as less important as a management strategy than do the other viability groups. As to the decision-support systems, more than 57 percent of respondents do not use and have no intention of using 'computer-based' decision support programmes, 24 percent make some use of them, and only 8 percent make extensive use of them. They were asked specifically about use of the National Agricultural Management System, discussed in Chapters 6 and 7, and there was almost no use of that. The system is currently suspended, but participants were asked if they had ever used it and, given the age of the respondents and the fact that NAMS started in 2006, there should have been some opportunity for exposure to it.

With regard to what the respondents would like as additional scientific information for drought management (Table 12.2), the main item seems to be more accurate forecasts, though even here there is no overwhelming demand. There does not seem to be much demand for better long-term forecasts and there is little demand for agronomic advice; the respondents do not necessarily want to know more about government programmes.

Approximately 73 percent of respondents agree or strongly agree that governments should support farmers in preparing for droughts. On the usefulness of government programmes, the preference seems to be for the familiar current and past programmes involving transfers from government (Figure 12.3). Interest rate subsidies were one of the main forms of earlier drought assistance, followed by FMDs and family support payments (see Chapter 6). Surprisingly, stock feed subsidies, which are not available in all states (and there are no significant differences in responses from the different jurisdictions), still rated relatively high. Respondents from WA rate the usefulness of FMDs lower than those from other states, perhaps again a reflection of the run of drought years they have experienced since they would be of less use after a run of low-income years has depleted

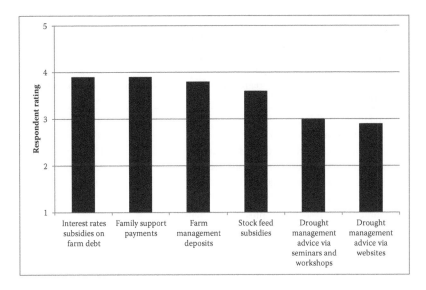

FIGURE 12.3
The usefulness of government drought assistance programs.

savings. There is some support for the advisory services, both online and through workshops.

More than 42 percent of respondents think governments should provide assistance to farmers during all droughts, while 53 percent think there should be assistance during 'only the most extreme' ones. Finally, respondents were asked about what information or evidence governments take account of when taking action on drought (Table 12.3). Media attention and political motivations are seen to rate very low. There is a relatively high proportion who 'don't know', followed by the conventional indicator of low rainfall. This suggests a lack of knowledge amongst some—perhaps even a lack of interest—and otherwise some presumption of attention to reasonably rational triggers such as rainfall deficits.

Discussion and Conclusions

In this volume, there are arguments for, and examples of, efforts both to improve drought science and to develop new ways to provide that knowledge to end users. The results from this survey raise some questions as to the extent and direction of those efforts. There is no great demand for different forms of science or for existing or new decision support systems. Expectations of medium-term forecasting seem to be relatively constrained, perhaps rationally so given the state of the science and the limitations of

TABLE 12.3

Perceptions of Policy and Program Drivers and Triggers

What information or evidence do you think the government takes account of when making decisions on which regions are eligible for drought relief?	Proportion of Respondents (%)
Don't know/nothing	30
Rainfall measurements/history	28
Referencing to its own agencies/departments	13
Consultation with local boards/councils	12
Analysis of natural/regional conditions	11
Income/productivity/profitability/yields	11
Community representation/pressure/lobbying	9
Ongoing hardship	7
Political motivations	5
Weather forecasting	3
Water availability/access	2
Feed availability	2
Media attention	1

probabilities as a guide to action, though there is interest in improving the accuracy of short-term forecasts, which would help the daily management of farm operations. There are four factors that should be considered as qualification to these tentative conclusions.

First, these are Australian farmers and, as argued earlier in this volume, the science here has been limited compared to that of the United States. Perhaps the respondents might have more specific demands for information and support systems if they had exposure to more options, though as noted in Chapter 11, low confidence in forecasting is a feature of farmers across different cultures (Ash et al. 2000; Jochec et al. 2001; Luseno et al. 2003; Coppock 2011). Second, there may be specific cultural and psychological characteristics at play and, in particular, Australian cultural agrarianism tends to have a streak of stoic fatalism running through it. The early settlers struggled with climate variability, rapidly declining fertility on some soils, vigorous woody vegetation regrowth and fires, which may set the pattern of thinking of the elements as forces that can be neither avoided nor turned aside.

There appears to be some implicit fatalism in the responses here, with seeming acceptance of the inevitability of drought, ameliorated by adaptive management as conditions worsen and government providing a safety net. Drought is overwhelmingly considered to be a natural event, which may, as discussed in Chapter 2, lead to a greater acceptance of the associated risks. On the other hand, some studies find that farmers in general believe they are learning from droughts and their management adapts accordingly (Saarinen 1966; Coppock 2011) and, more specifically in Australia, there is evidence that forecasting is being included in drought preparedness planning (CLIMAG 2001; Stone and Meinke 2007).

On the other hand, most respondents are not expecting climate change to make things worse. This may be a function of the wish to avoid cognitive dissonance (Chapter 2), in that reduced rainfall could be an existential threat—if not to the farm enterprise, at least to certain modes of production. Alternatively, climate change has strong political connotations and there may be some degree of rejecting the idea because of its association with particular groups and political parties. Another explanation, or perhaps contributing factor, could be scepticism about the sources of the scientific information about climate change (Chapters 1, 2, and 3), with the relevant scientists being seen as impractical boffins obsessed with abstract modelling and having limited understanding of what is happening 'on the ground'. Then there is perhaps an issue of time for acceptance, since 'old' risks are generally more acceptable than new ones (Botterill and Mazur 2004, 14) and occasional drought is an old risk, whereas climate change is a relatively new one. As noted in Chapter 3, rural people in Australia tend to be less convinced about climate change as an anthropogenic phenomenon than are others (Donnelly et al. 2011, 14), so there are at least some of these or other social, psychological, and/or economic factors at play in rural societies.

The third qualification is that this sample skews to the upper age brackets and is overwhelmingly male, to the extent that there are insufficient women to check for statistically significant differences on most survey items. The average age of Australian farmers is relatively high and the agricultural sector has a reputation for male dominance (Alston 1997; Grace and Lennie 1998), but women can be influential in decision making on some farms and responses to risk vary according to the sociodemographic characteristics (Po et al. 2003, 14; Ekberg 2007, 351). Finally, there were few differences by state, which was the only location variable. There may, however, be finer scale regional differences, resulting from a concentration of particular types of farms or other local cultural differences. Stone and Meinke (2007) point out that in regions where there has been successful uptake of more complex climate and weather information by communities, there are different rates of uptake of scientific and management advice.

Turning to the final purpose of the survey to explore attitudes to government policies and programmes, it seems that despite 30 years of trying to shift farmers and the general population to a mind-set of self-reliance and limited government support for businesses and individuals, the respondents to this survey still see government as having a safety net role. Molnar and Wu (1989) have postulated in relation to US agricultural policy that attitudes may change as policy and policy rhetoric change. This was obviously the hope of governments in emphasising preparatory rather than reactive management and policies. Despite that, the respondents to this survey still strongly support drought safety nets, which redistribute risk.

This is not to argue that farmers are alone in clinging to agrarian exceptionalism (Chapter 2) because survey work from 2009 suggests that there is considerable residual sympathy for farmers and even strong support amongst the general public that they should receive more assistance from government (Cockfield and Botterill forthcoming). What is surprising is that many respondents seem to be unsure about what leads to governments' triggering drought programmes, while a substantial minority think that it is just a matter of governments observing the rainfall deficit. This shows some unexpected faith in the rationality of decision makers and, for those who do not know about the triggers, at least a lack of cynicism. It has been argued in this volume that, while the physical manifestations of drought obviously influence policy triggers, political lobbying and media attention also have a role (Chapters 2, 4, and 7).

This survey, though modest in scope, reveals some expected outcomes in the respondents' caution about drought science, their support for government programmes, and their scepticism about the potential impacts of climate change on climatic trends. On the other hand, there are some surprises in that there is no great demand for new science or translational scientists to aid drought preparedness. There are few apparent differences amongst the farmer groups, though as noted, the location variable (states) is at a high level and young people and women are under-represented. The results are not robust enough to make a case for changing the current directions in drought science, information systems, or policies, but they do give pause for thought about the effectiveness of some research and policy.

References

Alston, M. 1997. Violence against women in a rural context. *Australian Social Work* 50 (1): 15–22.

Ash, A., P. O'Reagain, G. McKeon, and M. Stafford Smith. 2000. Managing climate variability in grazing enterprises: A case study of Dalrymple shire, north-eastern Australia. In *Applications of seasonal climate forecasting in agricultural and natural ecosystems: The Australian experience*, ed. G. L. Hammer, N. Nicholls, and C. Mitchell, 253–270. Dordrecht: Kluwer Academic.

Botterill, L. and N. Mazur. 2004. *Risk and risk perception: A literature review. A report for the Rural Industries Research and Development Corporation.* Canberra: RIRDC Publication No. 04/03.

CLIMAG. 2001. Newsletter of the Climate Variability in Agriculture R&D Program.

Cockfield, G., and L. C. Botterill. Forthcoming. Searching for signs of countrymindedness: A survey of attitudes to rural industries and people. *Australian Journal of Political Science.*

Coppock, D. L. 2011. Ranching and multi-year droughts in Utah: Production impact preparedness. *Rangeland Ecology and Management* doi 10.211/REM-D-10-00113.1.

Donnelly, D., R. Mercer, J. Dickson, and E. Wu. 2011. *Australia's farming future—Final market research report: Understanding behaviours, attitudes and preferences relating to climate change.* Sydney: Australian Government Department of Agriculture, Fisheries and Forestry.

Ekberg, M. 2007. The parameters of the risk society: A review and exploration. *Current Sociology* 55 (3): 343–366.

Grace, M., and J. Lennie. 1998. Constructing and reconstructing rural women in Australia: The politics of change, diversity and identity. *Sociologica Ruralis* 38 (3): 351–369.

Hennessy, K., R. Fawcett, D. Kirono, F. Mpelasoka, D. Jones, J. Bathols, P. Whetton, M. Stafford Smith, M. Howden, C. Mitchell, and N. Plummer. 2008. *An assessment of the impact of climate change on the nature and frequency of exceptional climatic events.* Canberra: Bureau of Meteorology and CSIRO.

Jochec, K. G., J. W. Mjelde, A. C. Lee, and J. R. Conner. 2001. Use of seasonal climate forecasts in rangeland-based livestock operations in west Texas. *Journal of Applied Meteorology* 40:1629–1639.

Kirono, D. G. C., D. M. Kent, K. J. Hennessy, and F. Mpelasoka. 2011. Characteristics of Australian droughts under enhanced greenhouse conditions: results from 14 global climate models. *The Journal of Arid Environments* 75:566–575.

Kirono, D. G. C., K. Hennessy, F. Mpelasoka, and D. Kent. 2011. *Approaches for generating climate change scenarios for use in drought projections—A review.* CAWCR technical report. The Centre for Australian Weather and Climate Research.

Luseno, W. K., J. G. McPeak, C. B. Barrett, P. D. Little, and G. Gebru. 2003. Assessing the value of climate forecast information for pastoralists: Evidence from southern Ethiopia and northern Kenya. *World Development* 31:1477–1494.

Molnar, J. J., and L. S. Wu. 1989. Agrarianism, family farming and support for state intervention in agriculture. *Rural Sociology* 54 (2): 227–245.

Patt, A., and C. Gwata. 2002. Effective seasonal climate forecast applications: Examining constraints for subsistence farmers in Zimbabwe. *Global Environmental Change* 12:185–195.

Po, M., J. Kaercher, and B. Nancarrow. 2003. Literature review of factors influencing public perceptions of water reuse. Perth: CSIRO Land and Water.

Renn, O. 2008. *Risk governance.* London: Earthscan Ltd.

Saarinen, T. E. 1966. *Perception of the drought hazard on the Great Plains.* Chicago: University of Chicago Press.

Slovic, P. 1987. Perception of risk. *Science* 236:280–285.

Stone, R. C., and H. Meinke. 2007. *Weather, climate, and farmers: An overview: Commission for Agricultural Meteorology Special Report.* Meteorological Applications. Geneva: United Nations' World Meteorological Organisation.

13

Drought, Risk Management, and Policy: Lessons from the Drought Science–Policy Interface

Linda Courtenay Botterill and Geoff Cockfield

CONTENTS

Scientists and policy makers in Australia and the United States face very similar challenges in the management of drought risk. Compared with other drought-affected countries, both have the relative luxury that drought is not accompanied by famine and is rarely a matter of life and death. Both nations seek to manage drought risk from within federal structures, which come with their own political, bureaucratic, and constitutional challenges. And both struggle with the difficulty, common to many liberal democracies, of managing the science/policy/citizen interface. This collection has sought to explore the issues that face both policy makers and scientists, including social scientists, associated with the uncertainty that drought, and in the longer-term climate change, bring to human activities that depend on water. We have brought together scientific experts, policy advisers, and contributors who have operated at the interface between science and policy in an attempt to unpack the complexities of developing societal responses to drought that are sustainable environmentally, socially, and economically. We believe that the mix of backgrounds and experience of our contributors provides a particularly rich perspective on these issues as it combines the findings of the theoretically informed research-based scholarly literature with the realities of the practitioner on the ground who seeks to influence and improve community capacity to deal with the phenomenon of drought.

As we stated at the outset, Australia and the United States are a good comparison because the many similarities between the two countries highlight the key differences, which include the existence in Australia of a National Drought Policy and the relative sophistication in the United States of drought monitoring activities. Beyond the narrow focus on drought, we have sought in this collection to address some of the challenges of evidence-based policy making and to highlight some key questions; first, around the relative importance of the diverse inputs into policy in democratic polities; second, around the nature of expertise; and third, around the challenges of understanding, communicating, and managing uncertainty. The following draws together some key themes from the various contributions to this volume.

The Nature of Policy Inputs

The recent interest in evidence-based policy suggests that democratic approaches to policy development, based in the contest of ideas between interests and societal groups, have not delivered optimal policy outcomes. The proposition is that policy processes would be greatly improved by strengthening reliance on strong evidence and moving away from decisions based on 'ideology'. Putting aside the many objections to this conception of the policy process as unrealistically linear, the idea of evidence-based policy leaves some important questions unanswered.

First, the origin of the evidence-based policy movement in medicine, where there are particular methodological approaches, has resulted in a bias toward evidence that is systematic and 'scientific' (in a positivist sense) over qualitative evidence or that collected through processes of consultation. Stehlik's chapter describes how the Australian government sought to broaden the evidence base in its review of the National Drought Policy by moving beyond economic analysis and climate forecasting and including an expert social panel to do two things: first, to synthesise the available social scientific research and, second, to conduct public consultations to gain firsthand information about the impact of the drought as lived experience. Stehlik's chapter contains a sense that this component of the drought review was a poor cousin to the other two inquiries conducted by the Productivity Commission and the Bureau of Meteorology/CSIRO. At the time of writing, it was not clear that this pessimism about government's attention to the findings of the consultations and the social science is warranted as the policy that will succeed the National Drought Policy (NDP) has not been announced. It is also worth noting that although the rhetoric of the NDP has consistently focussed on drought as a business risk to be managed by agricultural producers, the policy's implementation has reflected a much more diverse understanding of the impact of drought and has become increasingly focussed on the impact of drought on farm families. In the United States, sources of information about drought, such as the *Drought Impact Reporter*, reflect that the evidence that is accepted of what constitutes drought is much broader than just that provided by scientists monitoring and reporting on 'objective' meteorological, hydrological, and numerical economic indicators.

Second, the role of consultation in the policy process has been addressed in several of our chapters. The US Drought Policy Commission included 'listening sessions' in its deliberations, NIDIS is consulting with communities through its programmes, and the Australian Productivity Commission, as well as the Expert Social Panel, sought public input into their inquiries. The public policy literature recognises that consultation is an important part of the policy process but this does not resolve a central question of the relative status to be given to evidence from experts and that which emerges through these more democratic and inclusive processes of consultation. The evidence-based policy movement suggests that policy should be based on 'what works'—whether determined through pilot programmes or indicated by sound research. Critics of this approach argue that evidence-*informed* policy through which communities become better informed in their engagement with the democratic process is what is needed. This suggests a necessary triangular relationship between policy, science/expertise, and the community.

On one side of the triangle, the 'two worlds' of research and policy have some fundamental differences, and communication between the two is fraught—and the subject of a large and growing body of literature. Where uncertainty is a hallmark of scientific inquiry, policy makers seek certainty, or at least are prepared to 'satisfice' (Simon 1972, 168). Where policy makers need answers now, the research horizon is much further off—in some cases, decades away. The incentives that drive research are largely incompatible with the information needs of the policy process, with the peer review process in particular lengthening the process and imposing, perhaps quite reasonably, a degree of caution on researchers. Policy makers do not necessarily understand the peer review process and the academic's inclination to include caveats and footnotes in findings and recommendations. Conversely, the world of policy making can appear confusing and opaque to the outsider, the tight time frames are not appreciated, and there is limited recognition that much policy making is about decision making with limited information. Between the two worlds, there exist 'boundary riders' that understand both environments and we are fortunate that we have several of these individuals as contributors to our collection.

However, not all scientists have the time or the inclination to invest the considerable effort required to understand the policy process sufficiently in order to influence policy directly. An alternative route for scientists seeking to influence societal outcomes is the second side of our triangle—that is, communication directly with the community through education and community engagement. Our collection contains examples of how NIDIS and the NDMC have engaged in these activities either through establishing their own networks or through building on already established social capital in the form of existing civil society groups concerned about water management. This engagement brings its own challenges. As reported by McNutt et al., by Cockfield from the survey of Australian farmers, and as Stone and Cockfield note in Chapter 7, the availability of good scientific information does not always translate to its effective uptake. Cultural biases, such as the agrarian suspicion of expertise, and decision biases, such as the availability heuristic, can result in farmers and other managers falling back on other sources of information about drought. Climate change predictions about a hotter, drier future for both Australia and the United States have the potential to undermine the expertise of the scientists in the sense that predictions based on historical records may not be as reliable they have been in the past.

The third side of the triangle is, of course, the democratic process through which the community relates to and seeks to influence policy. This part of the relationship can be particularly problematic for experts who see 'nonobjective' considerations being introduced into policy consideration. In Australia, the National Drought Policy was designed to avoid the knee-jerk ad hoc responses that government had produced in the past in the face of nightly news footage of sun-baked, cracked earth, and dead and dying animals. The political resolve, however, did not survive ministerial council consideration of the proposed policy as Ministers agreed to incorporate measures to address 'severe' downturns. Less than 2 years into the new policy, in the grip of a deepening drought and in the face of a media-led appeal to raise money for drought-affected farm families, the government added a welfare component to the National Drought Policy. This element of the policy has come to dominate expenditure on drought relief, much to the disappointment of economists and other experts who have seen the policy change as 'backsliding' away from the rational risk-management approach. In the United States, the advanced monitoring tools developed by scientists are used by the federal government as triggers for the delivery of a plethora of assistance measures, many of which are inconsistent with the risk management message that is at the heart of the NDMC's mission.

The difficulties encountered in the relationships on all three sides of the triangle do not dissuade policy makers and scientists from seeking to engage and to ensure that policy is evidence informed. In the second part of the book we have a group of chapters that explore the various attempts that have been made to integrate different forms of knowledge into the drought policy process. Gene Whitney provides an eloquent and informed account of the US government's attempts to introduce a policy approach based on the risk management approach advocated by the scientific research community. His account illustrates how events elsewhere in the political stream can disrupt the policy process but also shows how, in the absence of policy action, the US government has supported financially the work of NIDIS in improving societal understanding of drought risk and promoting drought planning at local, state, and tribal levels.

Kerin and Botterill's account of attempts to raise the profile of the natural sciences in the development of drought policy in Australia, where economic perspectives have tended to dominate, illustrates how political will can be important in linking science with the policy process. Organisations such as the former BRS are important intermediaries at the science–policy interface as their employees understand both the imperatives of the policy process and the world of scientific research. The demise of the NAMS, however, illustrates the fragility of scientific monitoring systems that are set up to serve the policy process rather than providing the broader community engagement function that is one of the strengths of NIDIS.

The message from Section III of the book is that science can be made available but its take-up can become embroiled in politics. The example of the Apalachicola–Chattahoochee–Flint basin litigation and its impact on trust between community groups illustrates that even the science–community side of our triangle can become complicated by politics. And finally the survey conducted for Cockfield and Botterill suggests that outreach activities and decision support systems do not necessarily produce the expected results as farmers appear to prefer their advice to be delivered face to face and from trusted sources. As noted in Chapter 2, there is an argument that building trust in scientific institutions can reduce some of the frustrations of deliberative and participative policy processes, but this is a long-term exercise and trust is easily eroded.

Climate change modelling suggests that areas of Australia and the United States that are already drought prone will face hotter, drier conditions in the future. Governments, scientists, and communities will therefore continue to grapple with the challenges of negotiating a sensible drought response in a nonstationary environment. The authors in this collection agree that future drought responses should be based on an understanding that drought is a normal part of climate and that mitigation, risk management, and preparedness are preferable to ad hoc disaster responses. This approach will, however, require respect and effective communication between scientists, communities, and policy makers and at least some acknowledgement of the different types of experience and knowledge that each group brings to the task.

Reference

Simon, H. A. 1972. Theories of bounded rationality. In *Decision and organization*, ed. C. B. McGuire and R. Radner, 161–176. Amsterdam: North-Holland Publishing Company.

Index